GRIST TO THE MILL

GRIST TO THE MILL

Adventures in wheat breeding:
William Farrer, Nathan A. Cobb, F. B. Gutherie
and George Sutton

June Sutherland

A.D.2001

First published in 2001 by June Sutherland

Copyright © June Sutherland 2001

Cover photograph: S. T. Gill, December Watercolour, 21.8 x 18.2 cm, National Gallery of Australia.

Typeset by Chapter 8 Pty Ltd

Printed and bound by Ligare Pty Ltd

Australian Cataloguing-in-Publication Data
Sutherland June.
Grist to the Mill: adventures in wheat breeding: William Farrer, Nathan A.
Cobb, F. B. Guthrie, George Sutton.

Bibliography.
Includes index.
ISBN 1 86467 091 6.

1. Wheat — Breeding — Australia — History. I. Title.

633.1120994

CONTENTS

The cabbage, first in order of date in our kitchen-gardens, was held in high esteem by classic antiquity, next after the bean and, later the pea; but it goes much further back, so far indeed that no memories of its adoption remain. Hidstory pays but little attention to these details: it celebrates the battle-fields whereon we meet our death, it scorns to speak of the ploughed fields whereby we thrive: it knows the names of the kings' bastards, but it cannot tell us the origin of wheat. That is the way of human folly.

Jean Paul Fabre The life of caterpillars. 1916

And he gave it for his opinion that whoever could make two ears of corn, or two blades of grass, to grow upon a spot of ground where only one grew before, would deserve better of mankind, and do more essential service to his country, than the whole race of politicians put together.

Daniel Defoe, Gulliver's Travels

ACKNOWLEDGEMENTS

WILLIAM AND STEPHEN SUTHERLAND AND JULIA FREEMAN who have helped me do a professional job.

Dr Colin Wrigley, Dr Heinrich Gruasgruber, Dr David Chitwood and Donna Ellington for unexpected information, A.B.Blakeney.

Dorothy Blanchard and Dr Grace Iverson for family material about their grandfather, Nathan Cobb.

The Staff of the Agricultural Research Institute, Wagga Wagga, especially Marsha Reilly.

The deSalis family, Sosthene de Vilmorin, John Winterbottom and edward Duyker.

Charles Sturt University Riverina Archives.

And everybody in libraries for whom nothing was too much trouble.

MEASUREMENT CONVERSION

THE DOCUMENTS use the imperial weights and measures. Up to 1954 Australia continued to use these same weights and measures. The following is a conversion table for basic measurements of length, breadth, capacity and currency

12 inches =1 foot =30.48centimetres

1 mile = 1.61 kilometres

1 acre = 0.41 hectares

1 pint = 568 millilitres

1 gallon = 8 pints = 4.55 litres

1 bushel = 8 gallons (the weight of the bushel depended on the quality of the grain)

20 shillings = £1 (pound) approx $2.00

5/– (shillings) = the average daily wage for a workman

PROLOGUE

THE IMPORTANCE OF wheat to Australia has not diminished in the 20th century but the breeding of wheat has become a sophisticated and complex team operation, a long step from the last decade of the 19th century.

Over a century ago a group of men came together, almost by chance, who were to change Australia from a wheat importing country to an exporting country. The initial member of the group was William Farrer, who has been said to have done for the wheat industry what the Macarthurs did for the wool industry.

William Farrer had from 1886 worked on his theories of wheat breeding, but was enabled to achieve his results with the input of the others, men with other expertise, employed by the Department of Agriculture: Dr Nathan Cobb, F.B.Guthrie, George Lowe Sutton, James Pridham, and Robert Hurst.

The problems existed for which politically and economically, practical answers needed to be found but for a which a system did not yet exist. There were ideas within a fledgling agricultural department looking for outlets for solutions. The ideas which were expressed in a free-wheeling way through the media and private individuals needed to come together at the right time with a group of lateral thinkers who could coalesce the solutions to the problems confronting agriculture. It needed political muscle and convinced 'patrons'. Within the Department of Agriculture, the farmers organisations and Parliament conservative thinking made stumbling blocks, always advocating old methods for new problems.

This group with its input of creative thinking and practical application found their opportunities with the fledgling New South Wales Department, established in 1890 as an adjunct to the Department of Mines. New South Wales up to the 1880s was almost solely concerned with pastoral enterprises, but, pushed by newspapers and the new associations of farmers and settlers to provide guidance and education for the agriculturalists, the government of the day established the new department. In the financially straightened 90s it was to be threatened with being severely downgraded, if not abolished, and its funds decimated. It was in these tight circumstances that the work on wheat went on. The spur in this direction was the devastation of the 1888–9 harvest by the attack of rust in the wheat throughout all the Australian wheat-growing areas. The search for a cure was top priority in all the colonies of Eastern Australia.

William Farrer, Nathan Cobb and F, B, Guthrie, came together at the right time. At the beginning of their work they were effectively given carte-blanche to go ahead to attack the problem, G.L. Sutton, J Pridham and Robert Hurst in about 1900 added their skills. For nearly fifteen years they were virtually left to carry on without anybody really knowing what they were doing, whereas in 1908 the then free-standing Department decided that it was necessary to set up a supervising committee to determine what experiments were done and how they should be conducted. Albert Pugsley, a later wheat-breeder, said in a taped interview with Paul Ashton,

> *...they were more associated with individuals than to-day, when we regard institutions as having the responsibility of conducting agricultural research...so those years around the turn of the century were associated with individuals.*

EUROPE TRANSPLANTED

THE MANNER OF farming did not change in New South Wales to any marked degree for almost one hundred years after Governor Phillip had sown 8–10 acres of grain in what is now the Botanical Gardens

When the First Fleet was sent to the new colony or New South Wales, 21000 miles away across the globe it was expected that food could be grown to support the new occupants. Everywhere else a colony had been established this was so, and furthermore Joseph Banks had said it could be done.

Governor Arthur Phillip had not been provided with any of the real necessities of farming in the way of tools, the British supplies including poorly made hoes and axes. There were no ploughs until 1803.There were in the colony only a few who understood farming: the Rev Richard Johnson, Edward Dods, the Governor's servant, and James Ruse, who had been transported for burglary. In late 1789 the Governor sent Dods to establish Rose Hill[1] as a government farm of about 200 acres; James Ruse too was granted 30 acres there, as an experiment to see if a farm of that size could sustain a man and his family in self-sufficiency. The Reverend Richard Johnson who had a farm at Canterbury Vale[2] is recorded that here he grew wheat and citrus.

In 1790 Watkins Tench one of Phillip's officers recorded an interview with James Ruse, the time-expired convict to whom Governor Phillip granted the 30 acres[3] at Parramatta.

...I may proceed as fast as I can cultivate. I have now an acre and a half in bearded wheat, half an acre in maize and a small kitchen garden. On my wheat land I sowed three bushels of seed the produce of this country. I expect to reap about twelve or thirteen bushels...I sowed part of my wheat in May and part in June. That sown in May has thriven best...My land I prepared thus: having burnt the fallen timber off the ground, I dug in the ashes, then hoed it up, never doing more than eight, or perhaps nine rods a day; by which means it is not like the government farm, just scratched over, but properly done. Then I clod-moulded it, and dug in the grass and weeds. This I think almost equal to ploughing. Then I let it lie as long as I could exposed to air and sun; and, just before I sowed my seed, turned it all up afresh. When I shall have reaped my crop I propose to hoe it again, and harrow it fine, then sow it with turnip seed, which will mellow it and prepare it for the next year. My straw I mean to bury in pits, and throw in with it everything which I think will rot and turn to manure...My opinion of the soil on my farm is that it is good to middling; neither good or bad. I will be bound to make it do with the aid of manure, but without cattle it will be bound to fail.

James Ruse had been a farmer in his native Cornwall before he was transported for burglary and Tench's interview recorded the farming methods with which Ruse was familiar and which most settlers in the colony continued to practise in greater or lesser degree for almost the whole of the 19th century. For the poorer settlers unable to buy draft animals or the machinery at that time available, farming continued to be an extremely labour-intensive occupation and for them, and as it was for Samuel Shumack in the 1860s on the Limestone Plains, 30 acres would have been a large farm.

This failure of emancipist small farmers to thrive made a mockery of the British Government's aim of establishing in the colony a peasant society from the emancipated convicts. In Australia there was no evidence of indigenous peasant farming practice, as the Europeans knew it, as there had been in the Americas; it was the British agriculture which was transplanted.

For the pioneer agriculturalists in Australia it was literally starting from scratch, 'from the ground up' making improvements. To maintain the grant of land there was need for capital for stock and for draught animals and the manure which they produced. Rotation of crops to maintain fertility being rarely feasible on a 30 acre farm. The rentier, where his resources as the landowner could be used, would be more likely to make use of this practice, but if one can judge by the reading and the exhortations in the newspapers, many farmers would have scorned it.

From 1788 until the advent of the railways over the mountains, agriculture remained mainly within the Sydney Basin along the Hawkesbury River, north along the Hunter Valley and down the Southern Highlands. In 1804 Governor King sent a contingent to occupy Van Diemen's Land to forestall any French occupation. Agriculture quickly followed and when, between 1808 and 1815, the convicts and military were relocated to New Norfolk from Norfolk Island, labour was in good supply. New South Wales, by 1817 relied on the imports of wheat from Tasmania which, for many years, remained the major grain exporter to the mainland.

James Atkinson, one of the progressive farmers of the colony, in 1826 wrote a book of advice for prospective settlers of the state of agriculture in New South Wales:

> *I can give the average produce of the whole Colony from mere conjecture, but it does not probably exceed fifteen bushels per acre; and when the miserable system followed in cultivation is duly taken into consideration, its small amount is not surprising; on farms properly managed, the produce is about the same as upon lands of the same description in England. The smut in the wheat so prevalent in the Colony is entirely owing to bad husbandry. The same land is sown with wheat year after year without any change of seed, and without the smallest pains bestowed in preparing the seed.*

Atkinson, on his farm at Oldbury, prided himself on his expertise and good husbandry.

Australian soils are fragile and with the continual planting of the same crop year after year on the same land the earth became exhausted, the deterioration becoming more noticeable as the nineteenth century wore on. When the land was no longer productive the farmer cleared a new plot of land. Added to the decline in fertility were pests such as caterpillars and diseases such as rust, smut, bunt and 'take-all', to harass the farmer. It was Governor Macquarie, in the colony from 1809 to 1823, who remarked that there was 'a very great neglect of manuring and otherwise improving their land' and that 'it was too evident to ignore'. James Ruse had made virtually the same observation about the land he was farming at Parramatta-Rose Hill.

The earliest named varieties of wheat which were sown appear to have been English, such as Red Lammas, introduced by an officer of the garrison in Governor King's day. Others such as White Lammas, ripening in England at Lammastide, and creeping wheat[4], were the main varieties used in 1828. The Agricultural Society in 1822 imported seed from various countries but there was no visible differentiation in the seeds of the varieties being used, so that when the farmer sowed his seed broadcast 'the crop ripened unevenly and a great deal of grain was lost into the ground to the benefit of mice and parrots.'

There were men like Samuel Marsden and John Palmer. There were those who considered farming too arduous and labour-intensive, preferring the running of stock, as did the Macarthurs. John Blaxland said later:

We did not persevere much in agriculture because we could not manage it with so much other business, only by overseers, and under that management we found it a losing business.

Underlying this was an unfavourable cost structure highlighted by the contrast between profits from wheat and those from pastoral pursuits, marked by the difference in costs for labour and maintenance of that labour. Farm labour needed housing whereas, in general, pastoral workers could manage with their watchboxes or even with a lean-to of bark. There were rare large land-holders like Campbell at Duntroon who, on bringing shepherds from Scotland, built houses in the manner of a village for them.

After 1825 there was a lack of labour which hampered the larger wheat growers but favoured small wheat farms where the family were the labour force. In most people's eyes at that time 30 acres was considered reasonably large. The lack of mechanisation meant that farm operations were governed by the speed at which the work in the field could be done. Of the two tools used by the majority for most of the 19th century for harvesting, the sickle and the scythe, many chose the sickle. Although it was smaller and did not cut such a wide swathe, it was light enough to have been used in the old country by both men and women. However, in Australia, different circumstances in the early days of settlement set a pattern of division of labour. Dr Barrie Dyster in his book *Servant and Master*, wrote,

> *The agricultural and industrial revolutions in Britain sharpened the distinc-*
> *tion between men's jobs and women's jobs. Women on farms tended now to stay*
> *indoors and in the adjoining farmyard. Men monopolised the fields.*

The scythe used in farming operations as opposed to gardening was called a 'rake and cradle' which in the hands of good operator could cut a swathe which would all fall the same way. The resulting windrows would be left to dry in the field. With expert reapers, William Macarthur, could record 1837, that, at Camden, it took seven men fourteen days to complete the harvest of 31 acres of wheat.

The wheat acreage showed a steady expansion and by 1845 10000 acres of wheat was under wheat crop on the coast and 3600 acres were under crop on the South West Slopes, cultivated for the most by the tenant farmers of the squatters. This wheat was in the main for hay and the grain for the farmers own use. The extent of wheat planting increased, 10000 acres by 1860 and 30000acres in 1870. It was alleged that station owners after the 1861 Robertson Land Act, kept arable land a secret to deter the selectors claiming some of the acreage which proved to be successful wheat growing land.

Abolition of free grants in 1831 and the introduction of land sale by auction helped the Imperial Government's stated objective of 'concentration of owner-ship and the subsequent formation of a class of labourers for hire'. Prior to this land policy had been within the province of the Governors who had made free grants of land to settlers, discharged soldiers from the Napoleonic Wars and emancipated convicts as well as men acknowledged to have made substantial explorations. This created a problem as the land was rarely used for adequate food crops for the general population but for grazing purposes. Sydney was forced to import wheat and other products from the west coast of America with resultant high prices. In 1831 the Ripon Regulations were initiated whereby a minimum upset price of approximately a day's wages[5] per acre was set on the belief that 'industrious' immigrants would cultivate the land with profitable food crops. Over the years the upset price rose until it was four times that amount by 1842. This policy was always applied within the boundaries of the Nineteen Counties, but with more and more men looking for pasturage outside the boundaries some degree of regulation and record became necessary.

The Nineteen Counties, proclaimed in 1829, was an area of approximately 90,000 sq.km, from the Manning River in the north to Moruya in the south and to Wellington in the west. The rest of what is now New South Wales and Victoria were 'beyond the boundaries'. Beyond the Nineteen Counties the squatters were outside the limits of the law and its protection, but by the late 1830s the Government realised that it could not police beyond the 'limits of location'. Policing and military restrictions placed by the Government had not stopped the dispersion of flocks and herds so it was decreed that the squatters should have grazing licenses, renewable annually. This meant that until changes were made in the leasing policy in 1847, the licensees lacked security and thus were not inclined make a homestead and to establish family life, so making the government concerned about the prevailing immorality and lawlessness on these runs. The established policy prevented the establishment of towns and the sale of land but as soon as settlements were authorised agriculture and wheat growing developed in the neighbourhood. Those who bought land near the towns were not always people totally devoted to agricultural pursuit but what G.L.Buxton in *The Riverina 1851–1891*, called 'artisan-investors', such as happened in the vicinity of Albury.

Whatever the desires of the British Government about establishing flourishing agricultural pursuits, Governor Gipps was to say in the 1830s,

> *Australia, it must at once be observed is not an agricultural but a pastoral country, and dispersion is essential to its prosperity.*

The capitalists in NSW continued with animal husbandry especially when wool-growing became a profitable economic enterprise. The weather was less of a hazard than with agriculture, so with sheep the market expanded and export became profitable. On the other hand, for most of the 19th century the middleman (nearly always the miller), and not the farmer, was making the profit from the growing of grain. This continued until at the end of the century when the farmers developed a co-operative movement.

There was a differentiation of the land occupiers into graziers and agriculturalists. There was, with the exception of pastoralists who grew wheat for their own purposes, a concentration of wheat growing around the smaller settlements because of the economic forces of the availability of labour, transport and milling. Nearly every district had a flour mill powered by water or wind, to which the farmers bought their wheat to be stone ground in the way it had been done for centuries in Europe and on the same cost structure of part of the product being for the miller. Pastoralists grew wheat but for their own purposes. Sarah Mulgrave in her book *The Wayback* recounted the usual practice on her uncle's station which had been taken up in 1827:

> *The question of food was a pressing one in my uncle's early life at Burrangong and it was necessary to grow wheat for bread for the household, and oats and barley for the horses. The ploughing for these crops was done with a single-furrow plough, and the harrowing with an all-wooden harrow. The sowing, of course, was done by hand, and the crops were reaped with reaping hooks. The threshing was a tedious and a laborious work. At first the loose wheat stalks were put in bags which were tied up and beaten with sticks, but this method was too slow altogether. So my uncle brought and expert flailsman from Sydney and he, working all the year round, was able to thresh out sufficient wheat for flour and sufficient oats and barley for the stock. The wheat grain was sent to Goulburn to be ground into flour, until my uncle secured two steel hand mills from*

America…The grain was threshed on an earthen floor and then handpicked over and sieved and washed to clean it before it was ground. After being ground the flour was sieved to separate it from the bran. It was the duty of one man to grind the flour for the homestead and the workmen had to grind each Saturday enough flour to last their huts for a week.

In the colony of New South Wales, from Tasmania to Moreton Bay, there was a social dimension which marked farming as being second class; the small farmer was stigmatised by often being thought of as an ex-convict and as nearly every small farmers lacked capital, he could not raise sheep, he could not purchase more land and was compelled to devote his labour to tillage. This was often as a tenant farmer or working for a large landowner. Part of his indenture was being allowed some small acreage on which to farm. Samuel Shumack gave an account of his father being under just such a system with William Davis, the occupier of Ginnenderra on the Limestone Plains in the 1860s after the Robertson Land Act allowed selection. Shumack senior farmed about thirty acres successfully. Then he took up a selection on a portion of Davis' run which was open for selection. While he was still with Davis he and his sons set about improving it. William Davis was so sure they would fail and he would be able to buy back the land that he discharged Shumack Senior under the impression that he would come back to work for Davis. In the event this never happened. The Shumacks prospered and acquired more land.

For the small farmer wheat was the most remunerative crop because not only did it produce grain but there were other products, straw, hay and chaff. Maize, on the other hand, gave a greater grain return without the risk of the 'blight' but there was an advantage in wheat cultivation as there existed a market for all the products of the plant: wheat to be ground; hay, chaff and straw as well as the bran and other by-products of the milling process.

Wheat growing was the occupation of the small settlers not from choice, but from necessity. The perception of the lower status of farming saw many of those with ambition to rise drop out of wheat growing.

Australia was different from the old world as it had no peasant groups living in self-sufficient rural villages. Australian farmers had a different perspective of the objectives of agriculture: they were always market conscious. Their plan was to have, over their own needs, a surplus which they tried to sell for the best possible price. In the second half of the 19th century, as more settlers took up land further away from towns, more infrastructure of railways and grain storage was needed to make farming a marketable proposition. Distance to market was of far greater importance to the wheat farmer than to the wool grower. The wool could be many months on the road without damage, unlike wheat which, if it took the same length of time would be damaged by rain, weevils, birds and mice. Together with the problem of market accessibility, the expansion of the wheat area was hampered, often by ignorance of the condition of the land they held and of the use of viable farming practices.

To the English eye in the 19th century the practice of farming in NSW left much to be desired. It was severely criticised by English visitors because it did not have the orderliness of the neat hedgerows, the straight furrows between each headland, and the ploughman had often gone around the obstruction of a tree and continued on. The ploughing of a straight furrow was part of a long tradition of farming, a highly trained skill and not many of the settlers or the hired servants of the capitalists, tackling any pristine surface, undisturbed by any machinery since the beginning of time, could accomplish it. Agricultural Societies began to play a role in encouraging the acquisition of farming skills and ploughing matches became popular occasions wherever faming was carried on.

G.L. Buxton in his book, *The Riverina 1851–1891* described one Lutheran family's farming practice in southern Riverina on their well-established farm. When the Klemke children were about eight and nine in the 1880s they were important to the farm operations. Emilie Krause(nee Klemke) told Buxton in 1964, that she could remember that after her mother died in 1889 that she would drive the horses with the harrow. As their 600 acres of cultivated land was too extensive for the seed to be broadcast from a tub holding the seed hanging from one shoulder, the seats would be taken from the buggy and she, as a little girl sat in front to drive slowly while her father stood in the back with a bag of wheat would cast it out with both hands. By that time the Klemke family would have been farming since 1861–62 when the Lutherans had made the six weeks trek from South Australia to take up a selection in the Southern Riverina with other Lutheran families who had also come.

Broadcasting seed was taken as an art and farmers were judged on the skill with which the seed was broadcast. When the seed came through it was possible to make a judgement on a cold morning when the thin or uneven patches were shown up in the morning sun. Because of the critical nature of the broadcasting the farmer rarely entrusted it to the hired hand. The disadvantage of this method of sowing seed was that it was not done in rows and the farmer and his family had to beat off birds before the seed was covered with soil. Seed was still being done at the Wagga Experiment Farm in the late 1890s until the mechanical seed drill was introduced as part of the mechanisation of wealthier farmers.

RUST IN WHEAT

O F ALL THE DISEASES in the crops which affected the wheat growers most from the beginning was rust, already existing in Australia when the contingent from the First Fleet stepped ashore in Port Jackson. In all probability its host was what became known as Common Wheat Grass, *Agropyron scabrum*. Professor W.L.Waterhouse, the first Farrer Scholar, who became an authority on rust in wheat, quoted an extract from *The Memoirs of Joseph Holt* describing what would have been a common experience for the wheat grower.

> *On the 21st October, 1803, a more beautiful appearance of a successful harvest never flattered the expectations of a farmer; it was within three weeks of being ripe, the ears full and plump, the straw clear and well-coloured and in every respect it was gratifying to look at. In three days it was completely destroyed by rust and the produce of 265 acres was not worth £20. This extraordinary blight which I believe peculiar to this country is produced by fogs which come on suddenly and obscure the sky for some days, and if it happens when the wheat is nearly ripe, it inevitably destroys it. It covers the whole straw and ears with reddish powder, like rust of iron, which falls off as you walk through the standing corn*

In a book entitled *The Adventurous Memoirs of a Gold Diggeress. 1841–1909* by Mary Ann Tyler nee Brookesbank, and published by Kate Gibbs in 1985, Mary Ann described the effect of the rust on farmers at Airds, near Campbelltown, in her late childhood;

> *…wheat was one shilling a bushel. What beautiful wheat there was before the rust took over.*
> *We gave up our farm at Denfield. Rust took over the wheat completely.*

A little later she took up the story:

> *Every week father went with two loads of hay and two of his men would cart it with trucks. It looked really nice, the way it was stacked and had such a nice smell. New cut hay always has a fascination for me. But when the rust took over, the four men who had been helping father reap the 40 acres looked like red foxes–it was so rusty, and so the downfall of the farmer.*

She went on to comment on the loss of the farm:

I realised my parents were not well off since the rust took over the wheat. Farms were no good. Rents could not be paid.

Until later in the century the wheat varieties then being grown were always liable to rust. In the gradually expanded wheat areas west across the cooler slopes, one severe outbreak of rust is described by Samuel Shumack, in his autobiography, on its appearance on the Limestone Plains:

…Donald Cameron was on Gininderra, where William Davis had a paddock of wheat which was the admiration of the district. It was over six feet high and was as level as a table from one end to the other. It was a late crop and Davis had obtained the seed from America. He asked Cameron what he thought the yield would be and Cameron's reply was '50 t0 60 bushels to the acre at least if it is all like this end.'

'Come Donald, we will go through and see', Davis said. It was a hot day and both men wore white coats and although they were not short men the crop was inches over their heads. I have never seen its equal and I had to climb on to the fence to see over it. When Cameron and Davis emerged from the other end of the crop their clothing was red—this was their first experience with rust. A few days after the press reported rust in many parts of the State. This crop yielded sixteen bushels to the acre of pinched and milk-white grain.

The English wheat varieties suited the moister climates because of their longer growing period but in Australia they were always plagued by rust epidemics. Indeed by the 1870's wheat growing in the colonies looked insecure, as without fail they would be attacked by rust. It was becoming more profitable to plant other crops and then to import the grain. Farmers began to realise that other crops such as maize could be grown in those parts on the coastal strip which were subject to repeated rust attack. The drier parts of the colony, on the other hand were subject to hot summers which would wither the later maturing varieties, especially in New South Wales and South Australia.

New South Wales was backward in its approach to farming. Forward-looking farmers elsewhere were strongly advocating farming practices with intensive cultivation, regular rotation, heavy manuring and a dependence on high crop returns from small farms. Not only in advances in farming practice but New South Wales lagged behind Victoria by not opening up the inland over the Blue Mountains for farming until 30 years after Victoria opened up her interior. The mountain range was at the time the limiting factor in this matter until the advent of the railways.

In New South Wales, unlike Tasmania, Victoria and South Australia the lack of adequate transport hampered the extension of the wheat growing areas and wool became king. Long hauls lasting up to three months of wool from the Limestone Plains to the port did not present a problem as the wool would not deteriorate but wheat was always subject to the depredations of mice and weevils. The whaling, sealing, timber-getting and wool provided the income from export which enabled wheat to be imported into New South Wales. It was more economic to import from Tasmania and South Australia after its settlement, than it was to bring it from even somewhere like the Hunter Valley or over the Blue Mountains after it was settled. In the early days of the colony when drought and disease affected the crops, grain was imported from India Where wheat was grown in New South Wales, on stations and around the new towns it was only for local consumption. The difficulties with transport gradually became less for the Riverina, for instance, as the railway network expanded. By 1858 it had reached Campbelltown, by 1869 it reached Goulburn and by 1881 Albury was linked to Sydney.

FARRER TRANSPLANTED

W ILLIAM JAMES FARRER could not have expected to find both contrast and similarity between Lambrigg in Westmoreland where he was born in 1845 and Lambrigg outside Canberra where he died in1906. The shape of the landscape was similar but the sounds and the forest were an antipodean mirror. By the 1890s he could describe his new mountain landscape in as poetic terms as any of the Lake poets,

> *Here the clear and ever-running Murrumbidgee sparkles through our site and brightens it; here our beautiful Alps slackens the cold blasts from the west and south and refresh the eye with their ever-varying aspects; while the balmy but fresh and envigorating air gives such health and vigour as to cause work to be easy, enjoyment keen, appetite sharp and sleep certain.*

Westmoreland was bounded in the north by the border between Scotland and England and lay between the Irish Sea and the Pennines in West Yorkshire. The craggy Cumbrian mountain range enclosing the Lakes, attracted at the end of the 18th century painters and writers, who brought an intellectual quality to the county. Wordsworth and Coleridge were at separate periods editors of the *Westmoreland Gazette* using it as a vehicle to express their political views. The relatively sheltered more low-lying country north of Kendal was given over to farms where sheep and cattle grazed on the meadows beside the little rivers.

Kendal in the 19th century was a thriving manufacturing town built on industry since about the 13th century when leather and wool processing brought a prosperity to the town. Sheep were the mainstay of the farms round about. It had a charter from the 12th century to hold a market and later Queen Elizabeth had invited Flemish weavers to come and live and ply their trade. Beautiful Holy Trinity Church in Kendal built on the wool trade, and the nave was enlarged with donations from the Flemish weavers. In the parishes round about people wove cloth and an accounting of the amount produced by each village and hamlet was recorded officially. On some fast running streams woollen mills were a thriving operation.

William James Farrer came from a family of long-standing in the country between Kendal and Carlisle, bounded by the Cumbrian Mountains to the west and the Pennines to the east. For modern purposes, according to Burkes Landed Gentry, the family begins with William Fayrer in the 17th century. The horseshoes on the armorial of the family records a history associated with shoeing and the family motto 'Ferre va ferme', implies more than a comment on the original trade but reflects a more metaphorical attitude to life– 'iron goes hard'. William Fayrer is recorded as a member of one of the many Dissenting sects at the time, the Society of Friends, whose founder George Fox was one of the charismatic, itinerant preachers of the day. William Fayrer and his companions in 1669, at Appleby, refused to swear an oath of allegiance to King Charles II. For this they were sentenced to a term in Appleby Gaol.

The Quaker is next recorded as buying property near Greyrigg (about ten kilometres) north of Kendal. The fortunes of the family rose and fell as the members bought and lost other property but in 1839 William James Farrer's grandfather bought Wythmoor on Lambrigg Fell not far from Greyrigg. Not all the Farrers remained farmers nor even married. In 1806 Thomas Farrer's bachelor uncle, William, went off to Liverpool to become a successful grain merchant, having borrowed the necessary money from his sister Isabel. In 1833, Thomas Farrer's unmarried brother William followed his uncle to Liverpool to follow the same path, in due course taking over his business as grain merchant and becoming his heir. He himself built up an importing business in tea and coffee, becoming a man of considerable substance. His will sets out the extent of his wealth, estates both in Westmoreland and Yorkshire and considerable business interests in Liverpool, which when sold were to be made into a trust

This mixture of agriculture and business remained a way of life with the family down into the 20th century. The Farrers were a product of the industrial north, conservative and acquisitive; their business interests always related to agricultural produce and the acquisition of land. Thomas Farrer left all the acquisition of wealth to his brother, and if we can judge by his will, he was substantial but not wealthy. By all accounts William James Farrer, too, did not follow the family pattern about money until much later in life when he was concerned about his wife's circumstances after his death.

Thomas and Sarah Farrer, William James parents were married at Holy Trinity Church in Kendal in 1844, his father's occupation being given as that of farmer. Thomas had been born at Lambrigg and Sarah Farrer was a Brunskill who had been living in Lambrigg township, which included Lambrigg Fell and other sites which include Lambrigg in the name, such as Lambrigg Farm and Lambrigg Foote.

The places associated with the family all lie within a short distance of one another: Lambrigg; Docker where the children were born; Grayrigg where they were baptised and Oldfield where Thomas Farrer later farmed his 170 acres. Contained in Thomas Farrer's will written in 1875 is the extent of holdings and their location: Gateside in Howgill in Yorkshire; Croft Foote in Docker; Wythmoor in Lambrigg; a piece of ground in Ecclerigg in Westmoreland.

William James Farrer was born at Wythmoor, his grandfather's house, his name appearing in the Grayrigg Anglican church baptismal register as being born on the 3rd of April 1845 and baptised on the 9th of May of the same year, nine years into the reign of Queen Victoria and named for his great-grandfather

William. This same register records the birth of his sister Agnes in September 1846 and the baptism of his brother John in 1848. His brother Joseph was baptised here in 1851.

In the Census of 1851 Thomas Farrer is recorded as living at Oldfield farm and as having three servants: a male farm servant who at 47 may be accounted as being an experienced man about the farm; there were two 18 year-olds, a young man for the outside work and a girl who would have most probably attended to the dairy and helped Sarah in the running of the house. In the same census, only John is recorded as being in the house at the time, Joseph not yet born and it would seem that William and Agnes were away with relatives, their mother at that time being uncomfortably pregnant as well as being afflicted with the unnamed condition from which she died in 1852.

In the census Thomas Farrer was ranked as yeoman, *a freeholder under the rank of gentleman; a man owning and cultivating his own estate.* William Cobbett, 18–19th century political writer and reformer, distinguished a yeoman from a farmer in these terms; *Those only who rent are properly speaking farmers. Those who till their own land are yeomen.* This differentiation would have been in the minds of the British Government when it was making a review of land acquisition in New South Wales in the 1820s and 1830s. To the ordinary migrant looking around society at the time and at who was low and who was high these distinctions may have appeared irrelevant, labelling as they did Wentworth's ideas of a colonial aristocracy as a 'Bunyip aristocracy'. Farrer himself would have been conscious of these distinctions when he later came to New South Wales in both Sydney and on the Limestone Plains.

There is no record of Thomas Farrer having married again, but as his father at Wythmoor had died in 1851, it is possible the small bereaved family would have gone to live at there with his widowed mother and his sisters. One can speculate on who may have suggested that Farrer should be enrolled at the London school, Christ's Hospital. His home circumstances as a motherless child would have fulfilled the guiding principle of the institution. It may have been any one of the aunts or uncles who seem to have been education minded; perhaps the uncle who was later to leave Farrer an annuity in his will. There is however, no record of the circumstances of the enrolment 500 miles away from home.

The year after his mother died in 1852, William was sent to school in London, Christs Hospital, the Bluecoat School. The origin of the Bluecoat School was in the 16th century in the reign of Edward IV as a charitable institution to provide free or subsidised education for the children of the poor. The school was established in part of the erstwhile monastery of the Franciscans, the Grey Friars. When the monasteries were dissolved in the reign of Henry VIII the conventual buildings had become Christ's Hospital, the Bluecoat School

In Farrer's day, as it is to-day, the fees were based on an ability to pay, assessed according to the family income. The name itself derives from the long blue coat, the uniform from the 16th century founding. Although the school moved to Horsham in Sussex in 1902 it has maintained its links with the City of London and it continues to provide, as it did in Farrer's day, the same education of outstanding quality to children who would benefit from boarding in a caring environment. It gives preference as one must assume it did in earlier times to those who for some reason or other, whether it is financial, social or for any other reason which the school would judge as qualifying for entry. In this sense it appears as exclusive as any other English Public School.

When William Farrer went to the school in 1854, it was to a London which was the one which often became the setting for the novels of Charles Dickens. Although on the northern outskirts of London, 19th century London was crowding in; in Farrer's day, cattle would still be driven into that part of town to the Smithfield cattle market close by, where they would be slaughtered in the open. The smell and the noise would have permeated the school. At the tavern near the school, the Magpie and Stump, people would pay for good seats to watch the public executions in front to Newgate Prison within sight of the school. Across the field from the cattle market was St Bartholemew's Hospital.

At the school he excelled in mathematics for which he won Gold and Silver medals. In his senior years he became one of the School bigwigs, known as Grecians. John Middleton Murry in his book *Between Two Worlds*, recalled[6] his own days at the school and recounts the elevated status of these grand beings.

> *The Grecians in those days were visibly the lords of the School. When the masters were not actually teaching they disappeared from the school precincts. The domestic authority in the Hall was the Warden, in the wards the Matron—a being far inferior to the august Grecian, who listened to her conversation and her complaints with a distrait and preoccupied air. With his velvet cuffs, his multitudinous buttons(which all small boys believed to be stamped out of pure silver), with his coat of superfine cloth, and his gracefully drooping girdle, he sat on the polished granite stones, plumb in the middle of the Grecians cloister, itself plumb in the centre of the School. He was, indeed, the cynosure of every eye, the observed of all the observers; he was more, he was the navel and centre of the life of the School.*

His academic prowess carried him to Pembroke College in Cambridge, with an exhibition, a scholarship based on his academic record at school. Pembroke was one of the oldest colleges in Cambridge, having been founded in 1347 in the days of Edward III. The Foundress was Mary de St Pol the wife of the Duke of Pembroke.

In Farrer's time at Pembroke, for all its long history, it was still a very small College. The number of undergraduates admitted each year in the 1830s and 1840s was between nine and fifteen. This was not to change very much until 1871 when the statutes were altered to allow access to a degree which had been restricted to those belonging to the Anglican Church. Although small the

College had had as students men who had contributed to the knowledge of physics and astronomy, such as John Crouch Adams the discoverer of the planet Neptune. It was not without its critics. One such, in the year that Farrer graduated, remarked that Pembroke was *a delightful, drowsy, comfortable, venerable, useless caravanserai of idleness.*

It seems that in Farrer's time at the College, Pembroke was beginning to move away from the traditional core of Classics and mathematics into the natural sciences, history, law and other disciplines which today we would take for granted for a University to offer. He graduated from Cambridge with a double first in Languages and Mathematics, a Wrangler. Later on his French correspondent Henri de Vilmorin could recommend some articles on wheat in French publications knowing that he knew the language.

Farrer as an undergraduate joined the Cambridge University Volunteer Rifles and was commissioned as lieutenant, an almost natural progression from his status as Grecian at Christ's Hospital and a recognition o his family status. These endowments of authority would have given him the confidence to be at ease with the men whom he later met in Australia. In addition, he would have brought with him to New South Wales that place in the class system, which still regarded the Universities as the prerogative of the Anglican gentry.

After his graduation from Pembroke, there must have been family discussion about which career he should follow. The family wish was for him to follow law or the church, but he obstinately chose medicine, which suggests that in the time he was at school he was inspired, or influenced by St Bartholemew's Hospital to do some good for his fellow human beings, as he was to hold was his driving idea as he was involved with wheat. With apparently no real opposition from his family he began his medical studies, which in the event, only lasted a year. He was diagnosed as suffering from tuberculosis, something which in all probability had its start in his time in London and which had until the discovery of its cause in 1882, had a moral stigma.

THE REAL TURNING POINT

HIS UNIVERSITY FRIEND Frank Betts invited him to come to New South Wales to a new climate and new opportunities. It is most likely that the left England at the end of 1869 to avoid the winter. With an uncle and great-uncle as merchants and importers in Liverpool it is highly likely that arrangements would have been made for the two young men to sail on one of the clipper ships for a fast journey to Australia rounding the Cape of Good Hope. The first landfall after leaving Capetown was Cape Otway, marking the western end of Bass Strait. This was a treacherous coast and many ships came to their end when they missed the narrow passage through the strait. The run up the eastern seaboard to Port Jackson would have been done in good time.

Coming into Sydney through the Heads might have made the young William Farrer wonder where there was, in fact, an anchorage and a city to come to. To the north of the entrance he could see Manly a seaside holiday village of the day. That would have be barely reassuring but his friend Frank Betts, full of excitement and enthusiasm at his return to his home would have been able to be a guide to the sights as they continued up the waterway, past the forested shores with patches of clearing surrounding some quite substantial homes of the wealthier members of colonial society. The clipper would finally have been berthed at one of the wharves at Circular Quay close to the wool stores of the eastern side of the Cove, having threaded its way the many small vessels on the waters of the harbour.

Having arrived in Sydney, Farrer was welcomed into the Betts family at Parramatta. Not long after, Frank went on to New Zealand, where he was later drowned.

Pembroke had connections with Australia which provided links for William Farrer. The first Bishop of Calcutta, in whose diocese the new colony of New South Wales belonged, and later the first Bishop of Australia, were graduates of the College. Farrer's friend, Frank Betts, was the grandson of the Reverend Samuel Marsden, the chaplain of the Colony and responsible for the beginning of missionary work in the Western Pacific with its introduction into New Zealand. Besides the Betts connection, Farrer would have come with letters of introduction, and was to continue to be fortunate in the connections he established, whether deliberately or by chance. He would, no doubt, with his own dry sense of humour agreed to *Let us now praise famous men.*

In the Betts family there was an inherited interest in the wool industry in Australia, Samuel Marsden having been one of the pioneers of the industry in New South Wales. Farrer came with a new ambition, perhaps one he might have rejected in Westmoreland, to set up as a sheep farmer and it was through the Betts family that he would have been introduced into the milieu of graziers and sheep. While he was with the family in Parramatta, he apparently had a fall from a horse outside the Woolpack Inn, breaking his collarbone and it was said, affecting his sight. Although no one is sure of the precise date, some putting it just after he arrived, some putting it later in 1876, but Mrs Betts and her daughter Elizabeth nursed him in her home until he was recovered.

Farrer's intention to take up sheep farming at that time needed money if he was to be able to take up a conditional lease, as it would have needed to be stocked as well have the obligatory improvements. One must suppose that if he was given the money and the credit to begin this enterprise, the most likely person was his uncle, the Liverpool merchant. On the advice of old hands and his own observation, the way sheep were run in Australia and the husbanding of them was different from his own family experience. As ever, he wanted to learn all he could first and others had done before and would do afterwards, he needed colonial experience, found for him by his new friends.

In 1871 he travelled south to the Limestone Plains, travelling as far as Goulburn by train but the rest of the journey was by coach down a road which was not always free from bushrangers and broken axles. As far as Goulburn railway terminal, the settlers had to a large extent tamed the land to their own pattern, but beyond Goulburn there were still primeval forests where those who had deserted 'civil' society made their own tracks; men who preferred a rough and freer life. This part of the country in the decade before had been the bailiwick of marauding bushrangers like Ben Hall. The story would have been told and retold of the fight the young Faithfull boys from Springfield Station near Goulburn, had put up against the bushrangers. But things were inclined to be quieter now. Nevertheless the trip was not without its hazards of boggy roads and steep hills where the passengers had to get out and walk.

He had taken a position as tutor to the sons of Campbell, the Squire of Duntroon Station, where sheep were the prime occupation and all the surrounding stations were run in the same way. It was just what he wanted, considering his intentions and here he could find out what was needed for the venture to be a success. His status in the house would have been unusual because governesses and tutors were classed as servants and the university educated men who were coming to the colony and finding jobs as tutors were looked at askance because most were likely to be chronic alcoholics and unreliable.

The position was only for a relatively short length of time. In 1871 Campbell took his sons to England to be educated and he never returned to Australia. Farrer himself remained on the Limestone Plains in body and spirit for the rest of his life linked with the de Salis family. There has been speculation concerning Farrer's occupation in the years between 1871 and 1875 after he had left Duntroon. However, there are some signposts which can be followed. 1871 is fixed as is 1875 and 1873 in between. Four letters from Farrer to Reverend W.B.

Clarke provide four more dates which help make a chain of locations and activity; one from Queanbeyan in 1872 and three from Cobbity in 1873, where his connections with the de Salis family would have given him an entree to the houses of the Macarthurs and the Downes family of Brownlow Hill[7]

Farrer's relationship with the de Salis family began while he was with the Campbell's and taking part in one of usual entertainments afforded in the district Farrer met Leopold de Salis of Cuppacumbalong quite by accident. The young man was on his way to join a hunting party on Booroomba Station in the Brindabella Range when he became lost near Tharwa. Leopold De Salis encountered him and took him safely across the Murrumbidgee to join the campers. Farrer would have found in de Salis a man after his own heart who could discuss more than the price of sheep. Charlotte de Salis, writing about her Aunt and Uncle many years later could not be quite sure whether it was on this occasion that Farrer met Nina de Salis or earlier at Duntroon when she and her mother often called on Mrs Campbell. Henry de Salis, Nina's brother said it was at Cuppacombalong.

While he was at Duntroon he would have been an observant apprentice on the station, preparing himself for his planned enterprise. This was not to eventuate as he lost his money on a failed mining investment so he turned his mind to a less hazardous profession: surveying.

GRASS AND SHEEPFARMING

WILLIAM FARRER took an opportunity to impress his ideas on the bigwigs of the Colony when his time as tutor George Campbell's sons at Duntroon was at an end later in 1871, when the boys went with their parents to England for their education. He apparently spent the next two years as he said, *filling up my leisure time with writing a pamphlet.*

No doubt, as the result of his observations as well as discussions with some of the more knowledgeable people he had met, these 'men of mark', who were among the influential pastoralists of the colony, de Salis, Ryrie, Davis and Mc'Keahnie, he was prompted to write about the relationship between the goodness in grass and the soil on which it was grown. In doing this he could find an outlet for the academic disciplines he had acquired at Cambridge. There was, in addition, one other upon whom he came to rely for his knowledge and his opinions– the Reverend W. B. Clarke.

W.B.Clarke had begun his clerical vocation when he was ordained deacon in 1821 in England. He continued until he retired as the rector of St Thomas at North Sydney in 1871. His abiding interest was geology from the time he attended the lectures of Professor Adam Sedgewick at Cambridge. Before he came to Australia in 1838 he had presented several papers to the Proceedings of the Geological Society. In 1839 he became the headmaster of the Kings School at Parramatta. In 1846 he became the rector of St Thomas. His pastoral duties to his parish did not prevent him from actively surveying and collecting rocks and fossils. He discovered gold but on reporting to Governor Gipps he was told, *Put it away, Mr Clarke, or we shall all have our throats cut.* He was responsible, on the basis of his geological expertise, for locating several coal seams in New South Wales. It was to him that Farrer turned for expert information.

Farrer made the acquaintance of W.B. Clarke, who had made a detailed study of the Australian geology, and at the time of Farrer's acquaintance was the rector at St Thomas Church North Sydney. The young man had introduced himself to Clarke on a ferry crossing from the south side of the harbour, *as an old Cambridge man.* In a letter to Clarke he reminded him of continuing the conversation in the coach to Lavender Bay from the wharf. However, he had a purpose in thus writing to Clarke from Queanbeyan in 1872.The young man was seeking answers to some questions which he had about rock formations down beyond Cooma near Bibbenluke on the Monaro Plains.

I will ask you tell me what the mineralogic name of the particular trap(granite) rock which is so abundant on the Monaro Plains.

> *…what would you consider the difference in the main between these two classes of rock as regard, their chemical composition. Does one contain, more potash than the other and which? The trap I suppose contains more iron. Does either contain an appreciable quantity of soda or lime salts?*

This first letter written from Queanbeyan in 1872, like all the others he was to write to correspondents in Australia and abroad he sought information which would enhance and perhaps confirm his own knowledge and be one more step on the path he was exploring.

Clarke, in the course of a letter five months later, had written:

> *It might be a good exercise for your purpose to determine the composition of those rocks (on the Monaro), for which I for one would thank you.*

Over the first few months of the next year Farrer exchanged further letters with Clarke regarding him as a mentor and the most reliable source of information. In a letter from Cobbity written on the 7th of February he was anxious to have his information right:

> *Would a statement such as the following be sound? That 'the tableland of this colony almost entirely consists of Volcanic, Metamorphic and Paleozoic formations which with the exception of such alluvial tracts here and there, are better adapted for pastoral than agricultural pursuits.'*
>
> *If the statement needs qualification would you, be kind enough to point out to me the qualifications needed.*

There was no immediate reply to this letter so he anxiously wrote again on the 25th, reiterating his question, excusing himself for his persistence:

> *I would not trouble you but I don't know where I could get a book, which would give me the information I want. I am afraid that, as yet, you are my book.*

In April Farrer wrote once more to Clarke asking for another piece of information:

> *Is it a fact that all the permanent wheat growing districts of the old world lie in new stratified formations—mostly tertiary. I have in my mind when I ask the question the wheat growing district that lies on the N&E of Germany. In the E of England; the Nile country is of course alluvial. If this is the case what would appear likely to become the wheat districts of Australia, when we take into consideration the climatic and geological conditions. Should I be right in pointing to the S of S Australia, Western Victoria?*

In this letter he told Clarke about the pamphlet, the final outcome of his questions, and in a deprecatory tone concluded the letter,

My knowledge is too slight to cause the pamphlet to be of any value; but still I hope it will do no harm to any one. Its composition has given me employment.

When Clarke replied in July 1873, acknowledging the receipt of what he called Farrer's 'treatise' and had to tell him that the friend to whom had given it for an opinion was ill. He was reassuring to Farrer but advised him, *stick to the facts and let the theory go to the winds.* Farrer in later life would give this same advice to some of his 'apprentices'.

In 1873 William Maddock published William Farrer's pamphlet, *Grass and Sheep Farming.* When the pamphlet was first published, Farrer had qualified it, and in a sense absolved himself from any adverse criticism by calling it *speculative and suggestive.* In the first paragraph of preface he wrote a sentence which would have continued to hang in the air in later dealings with critics of his ideas:

Those who affect to despise theory will do well to recollect that a function of theory is to examine the foundations of practice, and by this means to modify and extend it advantageously.

He dealt with the idea of using speculation but continues with the view which he largely continued to maintain until the end of his life:

..I shall be thankful to anyone that may take the trouble to overthrow any speculation of mine that is unsound, for it would be showing me a faulty link in my chain of knowledge of the subject.

In spite of W.B. Clarke's advice to him about facts and theory, the preface discloses a young man who is in no doubt of his own abilities.

The following paper is written in much the same spirit as that which Montaigne says he wrote his Essays; the opinions, however that are advanced in it are not, I hope, what Montaigne says his ideas were–formless and undermined fancies. But I am sanguine enough to expect that they are of a more substantial character.

To liken himself to the great 16th century French essayist Montaigne was, in today's terms, to create an image calculated to impress his prospective readers: Besides maintaining his practical purpose he sought to confirm that he was not just an adept at chemistry but laid claim to the skills of literary composition exemplified by the great French essayist.

Even in 1873 he was not averse to putting on paper criticisms about some existing state of affairs. In the Appendix of *Grass and Sheepfarming* he wrote:

> *I believe there is or was a scientific Committee attached to the Agricultural Society: I have not yet noticed that it has ever taken in hand anything scientific; but I suggest that it should take the matter in hand: the Agricultural Society might grant some money for these analyses, if it has any idea of doing any good beyond holding shows;: there is just a possibility that Government might be induced to make a grant for the purpose.*

Nine years later a staff writer of *The Queenslander* had this to say about Farrer's paper, *a clever and most valuable pamphlet.*

In 1873, Farrer was more interested in engaging in sheep-farming, a vision of his future conditioned by his acquaintance with the Betts family descent from the Reverend Samuel Marsden, a pioneer sheep breeder, and by the graziers in the district between the Limestone Plains and Sydney. He saw a relationship between the use of the land, those who used it and the prosperity of the colony, particularly with reference to the Crown Lands Alienation and Occupation Act passed by the New South Wales Parliament in 1861.

> *…In the spirit in which the bill seems to have been passed, no consideration whatever appears to have been given to the true welfare of the colony, and the only aim of its framers to bid for the favour of the many voters at any cost, and with an utter disregard for the welfare of the colony! They have also tried to establish a state of affairs for which the colony is totally unsuited.*

The background to this criticism of the land laws was the evolution of land acquisition by grants in the colony from the 1820s, the purpose of which was to exclude the poorer settlers from acquiring land in order that they should become the land labourers. He had met on his arrival the Betts family connected with one of the most influential men in the early days of settlement, Samuel Marsden and he lived with, and among the bush aristocrats of the Limestone Plains. He met them socially, and determined to learn all he could, entered into discussions with them, all of them having extremely contentious opinions. In his pamphlet he obviously spoke from what he knew about men who had benefited from other men's land.

While Farrer was writing about this in 1873, 12 years after the Land Bill had passed into law, its contentious nature was not to be amended for another two years and for the rest of the century and into the 20th the land laws were to be amended as political pressures were brought to bear.

Farrer's censure was aimed at the lack of opportunity for men to develop and breed improved sheep. What he had to say about the problems of the Crown tenant, the pastoral lessee, reflected the frustrations of the flock-owners whom he knew and was familiar with. They had been forced to pay £1 and acre for land only worth five shillings an acre and then had to make improvement worth five shillings and acre, often only a slab hut fulfilling the requirement. If he had fenced before he had bought *he only offers an inducement for some may-be sheep-stealing selector to settle down in his paddock.* On the other hand, while not obliged to fence as an improvement, he could graze on another man's land which had been improved. He further presented an argument for achieving increased sheep numbers and greater wool production both by fencing and obtaining better and more nourishing pastures by ring-barking a certain proportion of trees. More sheep and more wool meant that more money could be assured, and so provide the Crown tenant could seek security of tenure. To reinforce this was the patriotic argument of *New South Wales …doing her fair share of helping … to put mutton and woollen goods within the reach of the poor at home.* Twenty years later his views were similar when breeding good wheat.

Beside the fencing in, the other improvement he considered a necessary improvement was ringbarking of trees, enabling more nourishing grass to grow, leading to bigger flocks and thus to the greater wealth of the colony.

Farrer himself had wanted to take up land but was disappointed in his ambition when he lost his money speculating in mining, possibly about the time he published the pamphlet in March 1873. 1875, the date of his surveyor's licence issue, was preceded by two years articled to a surveyor, so the time of his mining failure would have been in 1873. The cost of the travels he did prior to recording his observations of the geology of the area suggest that he was at that point in time still possessed of some funds.

Although Farrer was more interested in becoming a grazier, he may have read, at the beginning of 1873 a long article which appeared in January in the *Sydney Mail* concerning the views of Colonel Le Couteur about wheat growing and selecting of seed on the island of Jersey. On the same page there was advice of various methods of pickling wheat against rust. The work which Farrer was to undertake had forerunners of theories and practice which he came to consider more closely about ten years later.

Chapters of the pamphlet were published as papers in the journal of the Agricultural Society of New South Wales in 1877. By the time this happened he had gained wider experience of the variety of the land and vegetation as surveyor west of Dubbo.

Farrer, in *Grass and Sheepfarming,* balanced accepted practice against what existing scientific knowledge could be used to improve the value of the productivity of the land.

Although he sought expert information from W.B.Clarke his own education was further reinforced by his reading of the best that was available in the public and private libraries of the colony. His work exhibited an understanding of chemistry and the uses to which it could be applied. A modern scientist, reading the treatise, would remark that the analytical chemistry and our knowledge of plant and animal nutrition have advanced a long way, but regardless of these improvements in our knowledge, the principles which he outlined remain basically sound. For example, the mineral nutrient content of grass varies between species and within a specie between locations.

What Farrer had been able to work out for his pamphlet in 1873 had already been in a report which E.W. Hilgard, one of the pioneers of soil science in the United States, had prepared on the State of Mississippi in 1860. Hilgard wrote:

> *...my exploration of the state had shown me such an intimate connection between the natural vegetation and the varying chemical nature of the underlying strata that have contributed to soil formation, as greatly to encourage the belief that definite results could be eliminated from the careful discussion of a considerable number of analyses... to predict measurably the behaviour of soils in cultivation.*

Like Hilgard, Farrer was to continue to urge the importance of soils and their composition and promote agricultural chemistry, seeking to have scientific principles brought to bear on the investigations of the abundance or deficiency of the necessary salts in the soils. In 1873 he had written,

> *to what extent this is the case is unfortunately a scientific enquiry which has not yet received the attention it deserves, we are therefore obliged to be satisfied with the mere indication of results and with indirect evidence.*

He was to continue to publicise his views about the need for scientific investigation and agricultural education in New South Wales. After his death in 1906, Mr Downes of Brownlow Hill near Cobbity recalled making his acquaintance when Farrer was in the process of investigating the countryside.

Based on a particular analysis of the facts which he had assembled he could arrive at his conclusions. This is a characteristic of his later approach to presenting an argument or a result which he wants to be incontrovertible When it happened that he let an unfixed variety be tried and failed in the early days of his wheat breeding, he was angry with himself for doing it under outside pressure. Never again!

This paper on his observations and conclusions reveals the later man, even though he changed his direction and his views about land settlement.

SURVEYING

THE 1860s CREATED a problem. The selections after 1861 could be made before survey and identified with just land marks or blazes on trees defining the blocks. It was realised that with the growing number of people seeking to select, the boundaries would be inaccurate and lead to conflict between the selector and the pastoral lessee, as it later became the case in the Western Division. Deputy Surveyor General Adams in a paper to the New South Wales government stated that proper maps should be made using triangulation, which had been abandoned since the days of Major Mitchell. Victoria had proceeded with the task and covered that state with a complete survey. New South Wales had urgent need to survey the plain country west of the tablelands in the interest of settlement and agricultural development.

A difficulty complicated the proposal. There were not enough qualified surveyors and Adam proposed bringing qualified men from England. At the time when Farrer needed to find an occupation which would first of all keep him in the open air and allow him to use his previous knowledge of mathematics, the opportunity was open to him.

In 1875 he entered into articles with George Cummins, a surveyor based in Wagga Wagga who was working in the eastern Riverina. The Wagga Wagga Land Board covered the area between Tumut, Gundagai and the mountains to the east, down south to the Murray, west to Urana and almost to Narrandera; north to Temora and Cootamundra. In his two years as a trainee Farrer would have become well acquainted with the land and its settlers and squatters before he ever had anything to do with wheat. He may well have had time to apply the ideas he had expressed in his pamphlet, or even revise some his thoughts. He had little difficulty in passing the exams and on being granted his license he was contracted to the Lands Department to go to the Dubbo Land board district to work in the Western Division where he was based in the districts between Warren and Cobar

At the time he took up his duties on the Western Plains it was not to an easy task. It was recorded by one writer in 1854 that: *In summer the heat is oppressive and the country is frequently subjected to severe droughts.* For the traveller in the summer in the mid-century it was better to travel at night to avoid the heat and the flies. For the surveyor the winter was the best season for his work as the temperatures were not so oppressive and most importantly, there was a better chance of finding water after the winter rains. Charlotte de Salis in her recollections of Farrer said that he told his nieces he would feed his horses on bread because there was sometimes no feed.

The survey was a matter of measurement over distances. The early explorers had used the revolutions of their wagons wheels, or their horses paces, either of which supplemented the astronomical observations of latitude with the sextant. By the mid-century, the perambulator came into use, a wheeled device towed behind a horse or cart, but the accuracy of the survey was decided with a theodolite and chain.

For the surveyor the Western Plains surveying could be a hard boring job on what had been called the 'Dead Level', with the problem of survival looming large as the distances grew away from the settlements.

William Farrer wrote to Leopold de Salis 1875 July 11th at the beginning of his first licensed trip. The beginning of the letter was a request to enquire of Henry Parkes on behalf of Farrer's uncle about the purchase of a coal mine which Parkes had at Jervis Bay on the South Coast of New South Wales. The other half of the letter leaves no doubt about his self confidence in his ability to manage the task:

> *On my arrival in Warren this afternoon I start for the Bogan back blocks without delay. I expect to be hard at work there for 4 months at the least. The country there is exceedingly dry and is well-nigh waterless: for that reason I fancy my new-chum assistants will soon get some roughish colonial experience: one of them is made of the right sort of stuff, but of the other I feel somewhat doubtful.*
>
> *Hoping that you and Mrs deSalis are quite well and with every kind regards to your daughter*
>
> I am, my dear Sir,
Very faithfully yours
William Farrer.

It was in the Bogan backblocks that he may have begun the work pattern which he would follow even after he retired from surveying in 1886. Farrer's assistant during 1884–86, T. Kidston wrote to Mrs Macindoe in answer to her questionnaire when she was researching a book about Farrer's life:

> *Many times I wondered why he, after all day in the field, would sit up until midnight and after, writing to settlers and foreigners who were interested in the things he was, but who like him gained no monetary gain from. it asked him one day why he did it and he said thoughtfully, 'I suppose I do this for the same reason you play cricket and tennis. It is my recreation. A surveyor has little real need for physical recreation'.*

He revealed too, that Farrer had been the subject of gossip in Warren. The word was that he must have been rather dirty as he never changed his shirt. The truth, said Kidston, was that he had all his shirts made to the same material and the same pattern with enough pockets to take all portable writing materials and whatever else he needed to carry. Kidston said that contrary to gossip he was the cleanest man he had ever known–in body and mind.

With exception of Kidston's recollections there is nothing which can throw much light on his life in the Western division. There are no existing letters either to newspapers or people saved from these years up until 1882.

Thomas Farrer died in July 1876 and in 1878 William Farrer resigned from surveying and returned to England, where family affairs needed attention. In the time he was at his home, there appears to have been discussion with his uncle about his returning to England. In his uncle's eyes the young Farrer would not have been very successful, either on his own behalf or on his uncle's. His uncle changed his will and instead of a fortune the ingrate was only left an annuity. Farrer returned to Australia in 1879 to the dry brown Warren district to take up his surveying contract in the antithesis of the green and frequent rain of his English home.

PROGRESS OF IDEAS

IN THE MIDDLE of the 19th century Patrick Shirreff was experimenting in Scotland and he recorded his methods in *Improvement of cereals and an essay on Wheat fly,* published in Edinburgh and London.

The management of his plots was probably the first systematic planting for the express experimental purposes of growing pure strains of wheat:

My comparative trial plot of wheat might be described thus: on a field cropped with wheat, named and unnamed varieties were grown in parallel pairs, from twelve to fifteen feet long, and from nine to twelve inches broad, with a footpath a yard wide, surrounding the whole plot... from time to time, notes were made regarding each kind such as their ripening, length of stem, etc. by such means, the new varieties could be more readily distinguished from the old, and twice-naming detected, as the effects of soil and seasons upon the different kinds approximated... Then, commencing on one side, the seeds were placed by the hand at a given thickness, and each variety covered with earth before another was planted. By proceeding in this manner, the seeds were placed at nearly equal depths and distances, and the different varieties kept from intermixing in the process of sowing

By 1862 he had had this to say about new varieties

Many people believe that some plants can be altered by skilful treatment but my experience had tended to show that there is no way of permanently improving a species but by a new variety. I support of the view of plant improving, garden-ers can point to hosts of new and improved varieties of fruits, vegetables and flowers, while to corroborate farmers can bring forward the Chevalier barley, Swede turnip, Italian rye-grass and alsike clover. To this principle of improve-ment the cereals form no exception; and the small amelierat6ion which they have undergone in this age of progress may fitly be attributed to the apathy of corn-growers in this department of agriculture

New varieties of the cereals can annually be obtained from three sources-from crossing, from natural sports and from foreign countries.

Shireff also wrote:

before commencing to cross consider what properties the new variety is wished to inher-it and fix upon such kinds as possess in the highest degree the desired properties.

That this explanation was a template for Farrer's ideas remains speculative, but his friendship with Dr Joseph Bancroft in the early 1880s leads to the conclusion that through him Farrer may have read the publication. Joseph Bancroft, a dedicated naturalist was studying in Edinburgh at the time Shireff was working. The closeness to Farrer's more developed ideas by the late 1880s suggests that it was at this time he had access to the publication.

Ideas and theories were debated sometimes with great heat, in the newspapers throughout 19th century Australia. There were morning and evening newspapers as well as weekly papers; every town had its own newspaper where battles were fought, in some cases with language which today would be taken as libellous and William Farrer was not averse to voicing his opinions and disagreeing with other newspaper correspondents.

By 1882 he was showing his interest in agriculture. In a letter written to *The Australasian,* from Cuppacombalong in February, he maintained that agricultural land should be let out at fair rents, with security of tenure to the Crown pastoral tenants.

These views were most probably conditioned by his surveying in the land west of Dubbo where he would have seen that irrigation was a necessary adjunct to agriculture as well as the fact that his father-in-law, Leopold de Salis had long held the view that irrigation was an absolute necessity in the country away from the rivers. He had seen in his contacts with the station owners and managers, the difficulty associated with the necessity to grow fodder crops for those living as he had been in the far west of New South Wales.

The following extract, which appeared on November 19th 1882 in *The Queenslander* was based on these experiences surveying in the Western division of NSW,

'Mr Farrer of the Dubbo district in New South Wales, whose letter on rust in wheat appears in another column, sends us the following which should prove instructive to those who have stock which requires artificial feeding; and all stock is in need of it at times':-   I would earnestly recommend to the attention of farmers in arid parts of the colony the white Egyptian corn. Out of several sorts of drought-standing forage plants that I introduced some four years ago from Southern California, the Egyptian corn yields by far the best results as drought-relating. The next best to it is Kennedy's Early Amber cane. The millet I found to do well in sandy ground, with moderate moisture. The Pearl millet yields, under favourable circumstances, a most imposing crop; but the Golden millet looks like making splendid hay. I believe all the above plants would be perennial in Queensland; such is the conclusion that, I believe. Dr Schomburgk of Adelaide, to whom I gave the bulk of the seed imported, has come to with regard to them.

In the same issue they published the following letter which sets out his thought at that time on wheat improvement, the first of his public statements of his theories. 19th November 1882 in *The Queenslander.*

I notice in a recent issue of the Queenslander that rust has again made its appearance in the wheat crops of the Darling Downs and that the cultivation of Indian varieties of wheat have failed to secure the immunity from rust that was hoped from it. I trust that no one will be led by this failure to doubt that wheat-growing can yet be established as one of the great industries of Queensland; but it will not be established until a variety of wheat has been secured that is suited to the conditions of your climate. I beg to submit the following suggestions in regard to obtaining such a variety. I will first of all point out that the probability that a strong analogy exists between the rust of wheat and the American blight of the apple. Careful selection has been brought to bear on the apple and has resulted in the securing of a large number of varieties that are either blight-proof or so little liable to blight as to be exceedingly valuable. Until similar careful selection is brought to bear on the wheat, I believe that little headway will be made with wheat growing in Queensland. The rust-resisting variety may be secured by a happy fluke, but I do not think you ought to rely on the chance of that.

The process that ought to be gone through I believe is substantially (as outlined by La Couteur and Major Hadley) as follows:- Let farmers who have rusty crops this year go through them carefully, and see if they can discover any heads that are free from rust. Such heads should be carefully watched, and plucked when ripe, and sent to the National Association, or to some private person who would interest himself in this matter. 2 Some of these sound heads would in all probability, produce grain with rust-resisting properties. I would suggest that the grain from these heads be mixed and sown under conditions that would invite the occurrence of rust, and that the sound heads from the resulting crop be again selected and saved. If this process is repeated a sufficient number of times, I think it more than likely that a number of rust-proof varieties will be secured; but they will have been chosen for ability to resist rust alone. 3 The next process will be the selection from the rust-proof varieties of sorts that are also valuable for their milling properties. I expect that at this stage much might be gained by artificially crossing from different rust proof varieties.

William Farrer,
Warren via Dubbo
New South Wales, 7th November.

On the 25th November in the same paper was a comment by 'Jumbuk', the resident agricultural correspondent to the paper.

I was much pleased to see the letter by Mr Farrer on the above subject[rust in wheat]in your last issue; not because my own crude ideas on the subject tended in the same direction as his valuable and highly practical suggestions, but also because I believe the writer to be identical with the author of a clever and most valuable pamphlet, published some eight or nine years ago in Sydney, on the

*cause of 'worm disease' in sheep. In that pamphlet(which had extensive circu-
lation amongst sheep owners in this colony) the writer traced the cause of the disease to certain geological formations; and subsequent experience has shown that,
while other theories have failed Mr Farrer's has stood the severest crucial test. I
my surmise is correct, the editors and readers of the Queenslander are to be con-
gratulated on the fact that a subject of such vast importance to the agricultural
interest of Queensland has been taken up by a gentleman who is not only a
sound reasoner but who can place his ideas before his readers with such clearness, intelligence and ability.*

Although Farrer had been writing letters to the newspaper, they were largely
like his 1873 pamphlet, *suggestive and speculative*. Letters to newspapers was an art-
form in 19th century Australia, and Farrer would have been encouraged, if encouraged he needed to be, by Leopold de Salis, whose own published letters included
opinions on such matters as irrigation, ring-barking and the weather. Long distance
disputes surfaced in the letter pages of the innumerable publications.

Previously in August 1882, Farrer, having put the wheat problem aside, was
writing from Warren to George de Salis about the propagation of roses by grafting and the relative values of different varieties of tomatoes, apples and pears.

*I notice that it is given as a tip in regard to grafting, to cut off the scions which
you intend to graft at this time–before the buds have begun to swell–and to lay
them horizontally in a box containing sand, …moist and cover them with sand
about a foot deep: the box to be put under cover. This keeps the scions back and
delays the swelling of the buds, while the moisture in the sand keeps them in good
condition The grafting is done directly the buds of the stock begin to break.*

On March 31st 1883 an editorial on 'Rust in wheat' appeared in *The
Australasian* which was to bring him back into the fray

*Although the discovery of a variety of wheat capable of ripening a perfect crop of
grain under every condition of soil and climate could not be called an inven-
tion, the idea of accomplishing the feat has quite a charm for certain minds of
mechanical turn. Hardly a season has passed during the last twenty years without announcement being mad of the discovery of some new 'rust-proof' varieties
in that cereal. Australian farmers have been told to look to India for wheats possessing the desired immunity and we are bound to admit that in that direction
these should be theoretically the best chance of obtaining what we all must so
ardently desire.*
*No sooner has wheat growing become well established in newly occupied
country than symptoms of rust begin to appear, in fact about five or six splendid yields the dreaded parasite often makes a clean sweep of promising crops,
and this ultimately leads to the abandonment of wheat growing in the locality*

affected and to the return of the land to pasturage. Such was the early history of wheat growing in New South Wales; such is the history of the results of wheat culture in many localities in the Australian colonies in which that industry has been extensively carried out at any time.

…rust. Modern science has laid bare the mutations of the microscopical fungus which is familiarly known as rust, but our increased stock of knowledge of its natural history has not even faintly suggested a method of preventing its attacks.

…The experience in all ages has gone to show that soil containing an excess of stereoaceous matter combined with moisture in such quantity as to induce rank and watery growth will assuredly lead up to the development of rust. The spores of the parasite are during the summer months floating in the air.

Something is being is done in South Australia in this direction and, at one time, both Queensland and Victoria were moving in a promising way. The thrift of the government of one former colony and the party-political animus in the latter have swept away the nuclei of establishments, which legitimately developed might have yielded incalculably valuable results to Australian farmers and farming

Farrer's theories and observations continued to be published in the columns of *The Australasian,* where it is likely he had more exposure for his ideas than in *The Queenslander.*.

In the following letter of May 5th 1883 to *The Australasian* from Warren dated 24th April, it is evident that he is anxious not to have his ideas misrepresented, even by his own fault:

In the letter of mine that you published my bad handwriting has been the cause of your making a misprint which has quite destroyed the meaning of what I wished to say. In the sentence 'I expect them also to call to their assistance all other means of making their crops good and healthy such as their sowing in land in which the growth is likely to be rank, deep tillage etc' the misprint occurs. The word 'their' should be 'thin'. Thin sowing in soil in which the growth is likely to be rank is desirable, also early sowing, in order that the cuticle of the straw may have become flinty and strong before the dangerous season comes on; deep tillage, especially in a dry climate; drainage too, in soil which is liable to be compacted by water lodging in it.

I do not think we can rejoice too much over the new process of making flour in America; not only because the flour made by it is richer in the most valuable food constituents, but because the flinty grain which suits it best is produced by those varieties which are least liable to rust.

PS. I trust the agriculture reporter that you have sent to America will have much to tell us about the 'new process' method of making flour. I am quite prepared to have many ideas of my own corrected by his putting things in their true relative light do. The information I possess has all been derived from American journals and they have doubtless, presented things from their own standpoint alone.

In this May 26th letter he was writing about the silica in the cuticle of the leaf being necessary for it to be rust proof. He discussed this seven or eight years later with Nathan Cobb:

I recommended that grain from rust proof 'shoots' should be cultivated 'under conditions that would invite the occurrence of rust in order that the plants that did not inherit the rust resistant power of the parent plants might be culled out'.

A rust proof wheat such as could expect could be received, would be rust proof to about the same extent as 'blight proof' apples are blight proof.

Before the 'new process' of making flour was invented we were not in the position to seek for a rust proof wheat for the flinty grain which was correlated with the flinty or rust proof wheat stalk was unsuitable for making into flour under the old process

The matter of flour and the millers was to be an ongoing motivation for his work.

Farrer was stirred by criticism of his theories, voiced in the newspaper, to pursue his line of thought quoting the President of the Californian State Agricultural Society:

To improve the quality of seed should be the study and; about of some one or some class of men who could devote themselves exclusively to that operation...what I propose is that by a careful application of the law of selection, seed should be supplied to agriculturalists... from which seed enough would be raised to plant in their own fields and gardens. Mr A F. Blount, professor of agriculture in the State College of Colorado has produced some most astounding results by experimenting in this direction(work related to corn).

He wrote about the role of experiment farms with relation to the matter of good seed, especially in Queensland. He was take up the topic of the value of experiment farms in the improvement of agriculture about twenty years later:

I would suggest also in the task of securing rust-proof varieties ...to be undertaken on these farms.

After he had been experimenting with wheat and other crops at Lambrigg after 1886 and made his observations about the continuing quality of the land and how it was being maintained on his farm, he wrote to *The Australasian*:

I have often wondered that crimson clover(trifolium Incarnatum) has not come into greater use in this country. I have now grown it for three years, and find that our climate seems to suit it to perfection. it comes in early in the spring, is ready to cut for hay.

His interest in leguminous plants on the farm remained longstanding and later the chemist Guthrie was to institute experiments to test the value of leguminous plants and his department was to recruit a bacteriologist to experiment more precisely.

PROFESSOR BLOUNT

IN HIS LETTER to the 1890 Rust-in-Wheat Conference, Farrer quoted A. K. Blount as his authority. He had been corresponding with him about his wheat experiments since 1884.

From the 1870s the State of Colorado had run an agricultural college financed by state taxes. It was run by a board which was fundamentally interested in a 'scientifically-controlled system of farming'[8], looking for the best kinds of agricultural practice suited to the varied topography, mountains, plateaus and plains. The Board were directly involved with experimental work by collecting as many varieties of Colorado-grown grain as could be done. In 1879 they appointed Ainsworth K. Blount to be the farm manager to give the experimental work a scientific direction.   Blount's parents were New England missionaries who came to the west to work with the Cherokee Indians. He had been educated at Dartmouth College, and at the outbreak of the Civil War enlisted in the Union Army. After the war he returned to his studies in education and became involved with agricultural experimental research which was recognised by the U.S. Department of Agriculture. Over time he acquired a reputation as an important figure in agricultural grains research.

In the first of the existing letters to Farrer he wrote on February 25th:

…Your letter of January 25 together with copies of the Queenslander and the Australasian arrived yesterday. I have become very much interested in your account of rust etc. I am quite sure there is a way to prevent the ravages on the coast in good milling wheats. We are troubled with rust o soft varieties especially when sown on low alluvial soils in this altitude.

By a well established system of selection, crossing and selection again, I have made forty varieties that do not rust in our soils and climate, and some of them I am sure will not rust in yours. In alluvial soils, where a rapid and rank growth of straw is made, rust is pretty sure too prevail—in fact, where moisture, showers and hot suns come and go, such soils naturally invite rust. I have used common salt as a preventative. To be quite effective, one to five bushels of refuse salt or any salt sown broadcast upon a growing crop of wheat, oats, rye or barley will not only become a valuable fertiliser but will give the straw a siliceous coating capable in most cases of resisting the attacks of rust, particularly upon the stalk.

I have never recommended growing wheat on low land. High sandy loam with clay subsoil will raise the most perfect grain. Shallow plowing for wheat is better than deep. A harrow dragged over the growing crop when 4 inches high or the drill draw through the rows in the same way it was sown will facilitate the growth and ripening so greatly that more grain and better will be made—enough, more, sometimes to pay four times the expense of dragging or cultivating by drill.

In my report sent you by this mail it was impossible to sate the minutiae of the experimental work. I have only hinted at some of the most important steps to be taken in the work with each grain. I wish I could visit you. I think a four or six years breeding of your standard wheats would almost entirely revolutionise them so that rust, smuts, mildew and blight would be unknown. I came here six years ago. The soft wheats only were used for flour, your white Australian being the principal one. It then contained 11 percent gluten, a fair milling wheat by the way. Today it has become so deteriorated by non-selection that it has but 7 percent dry gluten and is not used by anyone except for feed. All the mills were adapted to grinding soft wheat. By my report you will see I introduced the hard wheats and commercial breeding up the standard soft and others received from all over the world. Today not a bushel of soft wheat is wanted in the mills and but little of it is raised. The hard varieties I introduced made by crossing and improved by selection are used all over our state, and the flour is better than Minnesota's best. It is called for by any large city and by South America and Mexico. I do not claim all the honour of this revolution—only a little. Our climate soils and facilities for irrigation are wonderful agents in producing promising fine grain When the mills found the flour made from hard varieties so much better they made changes in their machinery to meet the requirements imposed by hard wheats. So now our flour stands not only first in the Chicago market but our wheats stand first in the world(by the analysis of the 1884 crop) in milling properties.

I am not able to send you a copy of my report— the edition being exhausted. When the next edition is out I shall comply with your request. I will send you two copies one of which please send to Dr Bancroft.

The papers you sent me. I have read them all (except the stories and the ads) very carefully and shall keep them on file for the College.

Another thing I notice in your letter. Time of sowing. Wheat will germinate at a very low degree(17°). Now if you should sow your spring varieties in February, my impression is it would escape much of the rust. I sow wheat every year when the ground is frozen sometime in December or January. I find by doing this I can turn fall wheat into spring wheat at will. I will send you several kinds of my crosses and others to test. Make a thorough test of them in low land and high, wet and dry, under all conditions.

The three wheats you sent me through Professor Hilgard I sowed in October. They came up nicely and are now growing under the snow. I have great hopes of making quite a change in them. Crossing any foreign wheats on ours is unsuccessful but crossing ours upon them is always successful. I cannot explain this.

I have published no books. I shall be glad to receive the report or articles you speak of. I think there must be a wide field in your section for such work as I am doing here. From What I learn of your country is certainly one to which the world may look for finer grain in the future.

I am looking forward to a good deal of pleasure in our correspondence. Let me know all your best and worst pros and cons.

Thanking you for your long interesting letter and the papers.

I am truly yours

A.K. Blount

Eighteen months later Blount answered another letter of Farrer's:

...I must beg your pardon for not sending the spoke of in my last. When I came to look over my large assortment I found but few of value and I concluded to wait until harvest as I could then not only send you some nice samples but a product of all you sent me. I am quite busy harvesting and as soon as I get my grain threshed I will forward some fine samples.

My wheats are good, oats, barley, and rye fine. All the samples of wheat you sent me I sowed last September and have now a fair crop from all but three which failed to germinate a single kernel.

I have crosses of mine that will make 50 bushels per acre in field culture with only 30 bushels of seed. On my experimental plots I sowed 343 kinds of wheat, 45 of oats, 30 of barley and 30 of rye in rows 16 inches apart and one grain every 6 inches in the rows, making only 413 grains(kernels) to the square rod or about 6 lbs per acre. My largest grained wheat weight 1oz to 550 grains, the smallest 760 grains to make an ounce. The average is 530 grains and as I put only 423 grains on a square rod the sowing is 6.3 lbs per acre. Many of the kernels tillered so as to make 81 good heads from a single grain

To look at the crop before cutting it was as sowing. I shall report the whole when threshed and weighed up.

I hope to receive from you small quantities of your forage plants(seeds). The furze you sent failed to germinate a single seed.

Very truly yours

A.K. Blount

In 1887 an Act of Congress, the Hatch Act gave an impetus to research activity. The State Board of Agriculture insisted that the College's farm be managed experimentally as an instructional facility for students and as a means of improving the science of agriculture. Blount was said to have been a one-man band, full of enthusiasm for the task, not easy to work with. The range of his wheat experiments was enlarged: in 1883 he was experimenting with 221 varieties and in 1886 he had 343.

In his experimental work his most notable achievement was developing Defiance, about which Farrer and Nathan Cobb had a difference of opinion about its origin. Supposedly the grain originated with C. P. Pringle of the

Champlain Valley in Vermont in 1871. He gave Blount a small sample in 1879. Starting with heads not more than three inches long and containing on average about 21 kernels, he followed his practice of crossing and selecting to improve the variety. By 1885, he had plants which produced heads of five to six inches with an average of 42 kernels per head. At this point he could distribute seed to farmers who were farming under irrigation. By the 1890s it was the dominant spring wheat in Colorado and like Federation in Australia later, filled the State's coffers up until World War I

In a letter written in 1889 Blount wrote:

…Since you last wrote me our College grounds have been made an experiment station and the central government gives me 15000 dollars a year for experimental work. My work is doubled and much more extensive. The whole farm(240 acres) has become a field of plots for experimental work. My work extends over nearly 200 acres of the farm, 60 acres in pasture,20 in alfalfa. 20 in clover and grass,10 in corn, 4 in wheat 16 in _ acre and _ acre plots for special grasses…

You speak about sowing wheat on black alluvial soil—just the place to make rust. Such soils makes too rapid a growth hence a great deal of rust. The sap prevents the hardening of the stalk and the result is liability to disease. Now if you have high sandy dry soils out your wheat in them and when well up (4 9nches high) sow salt on it and note the effect and the difference. Put at the rate of 2 to 7 bushels per acre.

The grains of different wheats are of all different colours and shades. White, red, amber; transparency is a sign of hard wheat, not necessarily of gluten. Gluten …changes very much If not kept by selection a wheat will 'run-out'– that is the gluten will become so poor and small in the grain that it is fails to make good flour.

Large white wheats contain too much starch generally and have too little albuminoid. When wheats are sown in alluvial soils and have too much moisture(rain) the grain becomes inferior. Damp, mucky and foggy weather have the same effect and when followed by hot suns, rust is sure to be the result. I shall appreciate and make careful experiments with all the seed you send. At my next harvest in August I shall have many new seeds to send you. I shall begin to improve oats, barley by crossing this year.

Yours very truly,

A.K.Blount

DR JOSEPH BANCROFT

D R Joseph Bancroft became one of Farrer's original correspondents on the matter of wheat breeding. In 1890, at the conclusion of his letter to the delegates at the first Rust-in Wheat Conference held in Melbourne, he wrote: *Meanwhile I am enjoying the privilege of receiving valuable co-operation from Dr Bancroft of Brisbane.*

Both Bancroft and Farrer came from a similar background. Bancroft's father was a farmer and he was brought up there along with his three other siblings from his father's second marriage. He was given a liberal education and his first training was as a surveyor and then he began his medical studies at the Manchester Royal School of Medicine and Surgery, all the time maintaining his interest in nature and the countryside. He won prizes, not only for his medical studies but also for his pharmacological studies. In 1859 he graduated doctor of medicine from the University of St Andrews in Scotland.

When he returned to England to practise medicine he became president of the Nottingham Naturalists Society. A friend described him later as *utterly mad about natural history.*

Like Farrer, ill health forced him to leave England for Australia with his wife and family where he settled in Brisbane, setting up house and practice in Kelvin Grove. It was no long before his curiosity took his restless mind beyond the surgery door. Besides his immediate professional concerns he was interested in agriculture but had a mind which was never still. As well as investigating the causes of medical problems such as leprosy and typhoid, and all those associated with living in a tropical urban environment, he became interested in the pharmacological uses of native plants such as pituri.

From 1866 the theories he had, and the results of his investigations were made known through any outlet available to him: Queensland Medical Society, Government reports, The Royal Society of Queensland and the newspapers. The matters ranged from leprosy to kangaroos and echidnas; aboriginal food and poisons; indigenous medical plants and sanitary questions.

He turned his mind to practical matters in 1869 when he develop a process for preserving meat by drying it and after experimentation set up a commercial operation, The Australian Permican Concentrated Beef, on his property at Deception Bay. As long as cattle remained cheap the enterprise remained viable, continuing under the supervision if his son Dr Thomas Bancroft until the beginning of the 20th century.

Farrer and Bancroft began their correspondence about 1884. In that year Dr Bancroft had given an address on Indian wheats to the Royal Society of

Queensland which was reported in a paper such as *The Queenslander*, and it would have excited Farrer's curiosity, leading him to write to Bancroft. In February 1885 in reply to a letter of enquiry and some samples of wheat from Farrer, Bancroft wrote: *I will send you a paper on the Indian wheat question.* In the same letter Bancroft told him that he was not *answering any questions…about leguminous plants at present* but he enclosed some seeds of Maltese clover. In a letter to Farrer in 1885, he mentioned receiving seeds from Farrer's doctor friend in Georgia and his doctor friend in California. The passage of letters continued, and if Farrer's later correspondence is to be any yardstick, full of his ideas and details of what he had been reading, while Bancroft's letter were in a more telegraphic style of a man to whom time was precious.

Bancroft carried his background of farming into Queensland and he maintained that the development of the State depended on well-established rural industries with an urgent need of better farming methods and suitable crops. He wrote in about 1888: *The failure to produce wheat gives little encouragement to carry on agriculture on the European model,* besides which there was no flour mill in Brisbane and most flour was imported. In his views he found a ready co-worker in Farrer with whom he exchanged ideas, observations and seeds.[9]

Bancroft himself had contributed to a report commissioned by the Queensland government published in 1876 in which he advised the government to seek wheat varieties form all over the world, especially from places such as India within the same climatic ranges as Queensland. He selected from wheats sent to him by his friend Dr Dymock and selected rust resistant plants which were subject to milling and he then distributed them. He became the local authority on wheat growing and was chosen in 1888, as George Sutton was later chosen in New South Wales, to write a chapter for the Queensland Government's guide book for intending settlers.

He belonged to the Acclimatisation Society of Queensland. His letters to Farrer record his own experiments with wheat, rice and pasture grasses. His knowledge and enthusiasm enabled him to be an adviser as well as an experimenter. It was said, *one could not be too sure whether he was a botanist or a plant industry promoter.* Besides his garden at Kelvin Grove he had 1600 acres at Deception Bay where he planted seeds and plants from all over the world, including wheat, rice and barley. As well as his meat-processing plant he had oyster beds where he studied the biology of the oyster and experimented with cultured pearls. This was all week-end work.

In 1886 Dr Bancroft had hybridised a new grape variety using 'Isabella' and a variety called 'Sweetwater' and he outlined what he had done during 1885 in a letter to Farrer in 1885. He said that he had *impregnated the Isabella grape with the European pollen of many celebrated sorts.* It was about this time that Farrer was looking to plant vines at Cuppacombalong and Bancroft was writing to him with advice on the project.

It was said of Bancroft that for all his interest in nature and his contributions to natural history, botany, was not seen to be of immediate use and held little interest for him as it appears to have done for Farrer.

In 1888 and 1889 Joseph Bancroft became the Queensland representative on a commission examining methods of eradicating the rabbit. This meant him travelling to Sydney where Rodd Island in Sydney Harbour was the site of the trials. He apparently stayed with the Farrer's and the de Salis' family at Cuppacombalong as a later letter from him appreciated the meeting. Everybody in the house would have a community of interest with Bancroft because of their pastoral interests in Queensland.

The following letters bear out Farrer contention that he and Bancroft were friends and colleagues of long standing.

In a letter dated May31, 1904 Farrer wrote to Sutton with regard to two letters which Sutton had received and sent on to Farrer. They contained what Farrer called 'misrepresentations' about Farrer claiming to ideas which were not originally his own. In his reply Farrer wrote:

…a tissue of misrepresentation–I do not say intentional– but none the less misrepresenting. Dr Bancroft was an active correspondent of mine–I have been to his place in Brisbane and he has been to see me at Tharwa. The work which was done by Dr Bancroft and the old Acclimatisation Society I know. Dr Bancroft was the first to make an effort to combat rust: and for the purpose introduced, as being likely to suit the Queensland climate, a number of Indian wheats which were obtained for him in India by his friend Dr Dymock who visited there. The number of Indian varieties A, B, C, and D, all of which I now have in my collection. He also introduced the common, black-bearded, woolly chaff variety which is found under different names over the whole of India and which I have since received direct from India under several names. This Indian wheat(macaroni) Dr Bancroft found to resist rust the best of all. I received it also from him and it is in my collection under the name of 'Bancroft'. Other wheats which were tried by Dr Bancroft and the Acclimatisation Society were 'Defiance' and 'Belatourka'. No attention was given to the gluten content and no hybridisation (crossing) was ever attempted. The first suggestion, in fact, that crossing should be made use of for the purpose of making rust-proof (rust-resisting varieties) came from my pen and appeared in the Australasian (a paper which I then and for several years afterwards continued to…in) of Nov 17 1882. My communication to that paper ran as follows.

'I notice in a recent issue of the Queenslander that rust has again made its appearance in the wheat-crop of the Darling downs and that the cultivation of Indian varieties has failed to secure the immunity from rust that was hoped for it. I trust that no one will be led to be led by this failure to doubt that wheat-growing can yet be established as one of the grand industries of Queensland; but it will not be established until a variety has been secured which is suited to the conditions of your climate.

'I beg to submit the following suggestions in regard to securing such a variety. I will first of all point out the strong possibility that an analogy exists between the rust of wheat and the American blight of the apple. Careful selection has been brought to bear on the apple and has resulted in the recovery of a large number of varieties that are either blight-proof or so little liable to the blight as to be exceedingly valuable. Until similar careful selection has been brought to bear on the wheat plant, I believe that little headway will be made with wheat growing in Queensland. A rust-resisting variety may be… of a happy fluke, but I do not think you ought to rely on the chance of that.

'The process which ought to be gone through I believe to be substantially as follows (1) Let the farmers who have rusty crops this year go through them carefully and see if they can discover heads(plants) which are free from rust. Such heads(plants) should be carefully watched and plucked(harvested) when ripe and sent to the to the National Association or to some private person who will interest himself in the matter. (2) Some of these sound heads(clean plants) would in all probability produce seed with rust-resisting properties. I would suggest that the seeds from these heads be mixed and sown under conditions(climatic, I mean) that would incite the occurrence of rust, and that the sound heads (clean plants) from the resulting crop be again selected and saved. If this process were repeated a sufficient number of times, I think it more than likely that a number of rust-proof varieties could be secured: but they will have to be chosen for ability to resist rust alone. (3) the next process will be the selection form the rust-proof varieties of sorts that are also valuable for their milling properties. I expect that at this stage much might be gained by artificially crossing the rust-proof varieties."

Of course, if I had to write this now I would word it differently: but the germ of my work is contained in the above, which I wrote in my tent one evening after a heavy day' work at surveying in a hot climate. I was the first(I believe) to practice the crossing of wheats in Australia and if not in the whole of Australia, elsewhere than in South Australia, and I am the first to have practised it with the definite wish of making rust-resisting varieties, increasing the gluten content and flour-strength, improving the milling quality and percentage of flour and having a special suitability for our climate. The wish of the old Acclimatisation Society and Dr. Bancroft— practically the whole of the work was done by Dr.B.— was confined to introducing and testing varieties which were likely to prove rust-resisting. I was in thorough touch and sympathy at the time with what was being done in Queensland and know what I am talking about and I can rely on my memory in this matter.

You are at full liberty to shew this to Mr McK.

I am, dear Sutton,

Faithfully yours,

W. Farrer

In the original letter in *The Queenslander* Farrer's parentheses which were in his copy do not appear.

The letters which still remain reveal the world-wide exchanges which were taking place in plant breeding and the progress in the work which resulted from these exchanges.

Feb 23, 1889
Dear Mr Farrer,

In case you should be including seeds for your American doctor friend I enclose two seeds he might try with some advantage– I think you grow the Setaria and the Panicum crasgalli is in your strains most likely.

I sent among the ears of wheat some with solid stems—perhaps two herewith is a form he may as well try.

The solid stemmed wheat the Baron speaks of is grown near Menindee on the Darling and if you wrote to Mr Sadlier who resides near that town he could give you a few seeds. I saw it growing there and it seems to be, if not the same, almost identical to my old Indian form. I got a few seeds to grow and compare but the crows spoiled my chance of comparison.

My wheat heads are all thrashed out so you will need to wait a full season– or I could send you packets of seeds– shall I do so? For your Californian correspondent. You did not send me Dr Watkins letter referred to.

I would like to try the Tijania aquatica–Canada rice–several times tried without success–according to the Baron to be found all over the United states.

All creeping forms of fodder or other grasses (all grasses are fodder in Australia in hard times) we can get are worth trying. Various seeds are excellent–Arenda Donax–there is a seed from Cooktown northern coast promises well. I must send you some fragment's by post. It grows also along the shore.

Believe me, Yours truly

The Baron referred to in the letter is Von Mueller who was at the Melbourne Botanical Gardens, developing an extensive collection of Australian plants and cataloguing them.

The next letter is dated July 10 the same year.

Dear Mr Farrer,

Yours of the 6th instant to hand with its interesting contents. A few grains of Blount's wheats may as well go in even if late and I can sow them at the outlet of my water scheme where they will have ample chance of growing through the dry months and we may see what they are like. Enclosed are a few seed packets–which please return as soon as possible. I regret not having Blount's catalogue. I will send you shortly seeds I have kept over which you may send to the Professor. There are eight sorts of Egyptian sent out by Vilmorins; has sent Nile Valley wheats–the other eight I can't say who has sent them. They look like Indians and may be the Director-General of Indian Agriculture to whom I wrote about a year ago has sent them. They came late but I have put in a little to see what can be ascertained of this form. Of these you shall send some to America, one is the grandest looking grain I have seen. I am now raising Japanese rice for my experiments at the big dam plots. I should have liked to have tried the Canada rice you sent me a year ago which failed. Is there any of this growing with your friends in Bowen? Returning to wheat: the Russian sort–bearded black–has large soft grain–not small and hard as you suppose. I encloses a few grains, a single plant grew in the same patch of a beardless form. It seeded well but was rusted somewhat. I must have sowed but I have lost record, anyhow will let you have it when it turns up. Luckily if one omits to tally the sowing it is to tell what they are when in ear.

I shall be glad to get in return these seeds, grains of all Blount's sorts of wheat, barley, rye etc. etc. for an experiment in May 1890, should I live so long. The my large new dam and the adjoining field of cultivation will be ready and I shall be able to do justice to the question. I regret you should suffer from the virus. I am now ill with Bronchitis but not dangerously.

With kind regards to Mrs Farrer,

I am yours truly,

J. Bancroft

I send by seed post the packages of eight sorts of Egyptian wheat–duplicate sample–numbers 1 to 6, one set for you to retain for your own experiments and forward the other to Professor Blount. I find I was mistaken about a second lot of wheat I supposed from India. It turned out it was a duplicate of the Egyptian aforesaid. Some of my pearling barley is shooting into ear–rather too early you will say. However there may be grain.

A package of cuttings of vines which I hope the post will not object to transmit contained vines—hybrid black—hybrid white—Goethe—Seedless Grape sultana?—perhaps the currant grape, not the European—Iona Dutch Hamburgh or Gros Coleman or Black Monarco, one vine with three names—a great grower.

I have yet another good grape I think Isabella and three other hybrids—If you care I will send them also. Did the Crapin grass live with you?

We are now getting heavy rain.

All my wheats are barley look very well. I find these are growing three tufts of the solid stemmed wheat from Albernabate(?)—Mr Larkins sort. Did you get any from him?

I found the seeds of grain stored in glass prune jars with the screw-tin-lid kept good as seed. All stoppered and corked bottles destroy the vitality of the seeds. This I regret as some of my former wheat I put up in this way I would like to reproduce but all so stored are dead.

I am, yours truly, J Bancroft

On January 23rd in 1890, Bancroft wrote a long letter to Farrer discussing fodder grasses and the problems of keeping his grain seed because of the weevils.

Dear Mr Farrer,

I am writing you on thicker paper than usual as I find from perusing all you letters, June 27th 1887 up to date (some of the earlier ones are lost) that ink showing through the leaf is a great defect in reading correspondence.

I have made note of the following matters Tijania aquatica not yet introduced.

I have an extensive area of rice this year—have had however to contend between floods and high tides and shall not make a great success of it on my mangrove land.

Have you the water couch yet established? I was thinking to send you more joints of Crapin and for you to keep a plant of it from year to year in a flower-pot in a protected hose so then you may put joints of it in from time to time—when good roots are established in your garden under the protection of trees the plant may stand through the winter—the same should be done with other rare grasses until established—of Dr Watkins' American grasses I have so treated as yet only the Texan grass has succeeded to come up from year to year in my garden—but the Tripsacum dactyloides is firmly established but I regret to say is not valuable as fodder—coarse cutting leaf. The Crapin is perfection—makes a woven mat 5 feet high. You can lie down on it like a spring bed—animals are very fond of it and give it a severe biting down but the roots are not destroyed— Its form is like a giant couch.

Concerning vines-

I can recommend two additional hybrids—four in all—my intention is to set out Isabella for roots and when two years old to use them for grafting on— putting the scion down beside the rooted plant with a crowbar. I purpose planting out an acre or two in this manner, so then when I have decided what vines I will grow, to graft as mentioned.

The Racer grape, I have set one out–I had a hybrid strawberry fruit a few weeks ago–I will send roots of it and other kinds again shortly.

I have not done justice to the gardening book you sent me but will return it at once if required—am obliged for it as it contains some valuable advice.

Concerning wheat I sent a few ears of those you let me have in July last, though late I was surprised to get good grain without rust. They were however in a place where no wheat had been grown before.

The other two, black bearded and brown bearded (I cannot follow the numbers) I have sent a few ears of. The weather was bad and much of my wheat went rusty. The barleys are all right being much earlier than wheat. Sadlier wheat was not solid and it rusted badly and I have lost it—may keep a few grains for observation. I must try in Toowoomba to get again the only solid stemmed wheat I know—my seed with ears bottles up are dead.

Steinwedel's I have not yet tried.

Weevils are a great plague here in preserving seed wheat and I am at my wits end to know what to do. I had two bins made weevil- proof in the wood—wee-

villy grain put in them I have subjected to carbonic acid gas made with acid and alkaline carbonates. Then I have blown smoke from a fire box bellows to fill the chest over and over again with most suffocating smoke—still more keep coming out of the grain—from the grubs no doubt developing.

Now I think to coat the grain with clay and dry on warm iron plates—of course one must not destroy the germinating property.

How are those silos made in Asia Minor and elsewhere. The Asiatics seem to have conquered the weevils somehow.

In India earth is mixed with grain it is said—but this does no good in my hands.

We are now getting very heavy rain in Brisbane. The wheat from Lambrigg is very like my wheat from Cabool, to look at both are identical. Anyhow we shall see later on. My thrashing machine and engine are being set up in a new shed near my dam and in future I hope to thrash out as soon as possible after the sheaves are dry—then maybe have the seed duly prepared for the following year—to kiln dry or consume all the rest as soon as possible.

With kind regards to Mrs Farrer,

J. Bancroft

The year before this letter was written, Bancroft had, in an article written for prospective wheat growers in the *Queensland Illustrated Guide*, published by the Queensland Government, setting out a preventative measure against pests in stored seed.

To store wheat for sowing, pack the grain in bags that hold a bushel or less, and place on wire netting or other framework in the upper part of a smoky kitchen, under the shingles. The bags should be opened and the contents examined once a month, in case weevils develop in the centre of the bags where the smoke does not reach.

Bancroft's letter writing style seems to be a product of a habit of writing notes as he deals with his patients and observations. In to-day's world he would have carried his electronic notebook to record his wide ranging interests. Farrer's letters are stuffed with information and advice and comment, often over many pages; he would today have waited until he could sit in front of his computer screen and send off email with many attachments.

NATHAN COBB

WHILE WILLIAM FARRER was writing his lengthy letters to his overseas contacts, and just as the 1889–90 crisis in wheat was about to turn agriculture around in Australia, and especially New South Wales, a man and his family arrived in Sydney who would in no small measure help to effect that change. He was Dr. Nathan Cobb and his background was as far apart from Farrer's as it was possible to be. In personality too they were sharp contrasts. Farrer was laconic and disinclined to talk, while Cobb was the quintessential demonstrator. They were alike however, in their sense of humour and their love of children.

Nathan Cobb, although later an American scientific leader in plant pathology, has a place in British Imperial scientific work. He was, along with Donald McAlpine, one of the first two professional pathologists appointed to any government anywhere within the British Empire.

Although he was only connected with the Department of Agriculture in New South Wales for 15 years he was influential in the way it should go from the point of view of research and succeeded in influencing the physical structures and the equipment needed to carry out research.

Nathan Cobb was a polymath. Not only was he a pathologist, a botanist and painter, but when he needed equipment he could build dark-rooms and cameras, laboratory apparatus and microscopes. Because of this direct practical approach to research and his commitment to whatever he was working on his endeavours were the scaffolding which helped to strengthen the early years of the Department of Agriculture in New South Wales.

He was born in the United States in 1859. His father had, at various times, been a millwright and carpenter, mill manager and farmer. It was here that Cobb learned his manual skills which he could turn to good use. He only went to school during the winter months, and at fourteen he worked as a farm labourer but he was able to become qualified as a school teacher. A later colleague writing about him in the first issue of the *Journal of Nematology* as having about him an aura of 'New England austerity' a product of his time and environment.

Cobb's innate ability to teach is evident from his later articles to the *Agricultural Gazette*. While still in his teens this ability was demonstrated after he bought his first microscope for $25 by mail order, when he was working for Charles Prouty who was to remain his friend and benefactor. For Prouty's children he demonstrated and explained the minuscule natural things that could be seen through the lens of his treasure, as he did later on a much more sophisticated instrument for the de Salis children at Lambrigg. Prouty was so impressed by his knowledge then and his enthusiasm for revealing the secrets of nature that he suggested sitting for the teacher's exam. Having passed with no difficulty, Nathan Cobb found himself at the ungraded school of Wire Village. It was not long before he was in charge of District 3 Grammar School in Spencer. He realised by now that he needed more education for himself, regardless of how much he observed and sought out information.

His ambition to go to Harvard was thwarted by lack of money but in 1878 he passed the entrance exam for the Worcester Free Institute, where he could pursue his interest in natural science. He wrote to his future brother-in-law Fred Proctor about his dilemma:

> *You know I lean naturally to the natural sciences, not the physical. I care more to investigate the form and structure of plants and animals than to look into their chemical composition; but as each branch of science is dependent upon every other it is necessary to have a good knowledge of physical science in general in order to succeed in the natural sciences. Now my knowledge of chemistry amounts to this mark:0 so I intend to study chemistry in the his humble beginnings to the highest professional attainments Technical school solely for the purpose of applying it to the natural sciences…It is divided into two parts viz: Part I Theoretical chemistry, Part II Inorganic chemistry. Now I want organic chemistry, what am I going to do? In point of expense the Technical School is the cheapest school I can enter the tuition being free to the youth of Worcester County. There are no schools making a speciality of natural sciences excepting the high priced colleges. I would rather take a thorough course in the natural sciences than any other, but lacking the means am obliges, for ought that I can see, to study mathematics and earn my living by the tripod, making the study of nature always a pastime for leisure hours, when I would much rather study nature always and let mathematics go to grass. What shall I do?*

What he did was take his three year course in chemistry at Worcester, encouraged by Alice Proctor. His graduation thesis, *Mathematical Crystallography*, was the theoretical groundwork for the later exact modifications he made to microscopes and cameras and all that pertained to them always remained his profession and his hobby in his ongoing researches.

In 1881 he married Alice. He had met her when he was fourteen and her brother Fred took him home for tea. When he had begun teaching at Wire Village he boarded with the Proctor's, while Alice trained as a teacher at the Worcester Normal School, and remained until he went to Worcester. Both Nathan and Alice had an interest in botany and she introduced him to sketching and watercolours. These skills became hereafter his way of recording his findings and where he had been. It was through Fred Proctor that he became friends with a well-known New England artist, J. D. Greenwood. He absorbed the cultivated attitudes which carried him smoothly from Spencer to Amherst and Germany; to Australia and back again to the United States.

Just as he had finished at Worcester and married, a position became vacant at Williston Seminary at Easthampton, to teach chemistry and drawing. After the first year of teaching chemistry and drawing he was given charge of all the natural science and in order to teach it more successfully he set about having the laboratory equipped with more stools, better work tables, and microscopes for the students.

Frieda Cobb Blanchard his daughter, writing about him.

Our father hearing that new teaching laboratories were being installed at Harvard took a trip to Cambridge and brought back ideas for setting up a biology laboratory at Williston.

Williston became one of the first secondary schools in the country, if not the first, to have compound and dissecting microscopes for each member of the class. To the prospectus of the School were added: zoology, physiology and botany *with laboratory practice.* For botany and geology there were excursions.

His ideas about teaching and how a school should be run became evident when he was acting manager of the Wagga Experiment Farm in 1897–8 in the first year of the enrolment of students.

While he was living at Easthampton and teaching, he set up a laboratory in his home as he did again when he came to Australia. He set about practical application of his chemistry training, making chemical analyses of water, paper, beer, and other things from teas to butter, soap to gunpowder as well as many other products for manufacturers in towns nearby. While all this would have been for the benefit of his pupils, his private consultancy would have enhanced the family income and no doubt helped build the fund for further study.

In the area around Easthampton there were a number of academic institutions which would have stimulated the academic progress which Cobb would have been seeking. He began to be published in 1885 and the first was *Elements of Chemistry.* In 1887 published *Catalog of plants growing wild within thirty miles of Amherst* and then the same year a second edition, *A List of plants growing wild within thirty miles of Amherst...* One can imagine that the compiling and illustration of this work would have been a pleasurable co-operation between the two botanisers. It was this first thesis that became his entree into his doctoral study.

He had set his mind on enrolling at John Hopkins University, established in 1876 and specialising in the natural sciences. However, because of a change in position of the President of Williston he was asked to stay another year, and so doing, he lost his chance to take up a fellowship at John Hopkins for which he had applied. The cut-off age was twenty-seven and his twenty-eighth birthday had fallen two months before the date when he would have taken up the fellowship. Nevertheless, he was not to be deterred in his pursuit of his ambition. He sent off an application, in German, to go to Jena and study under Ernst Haeckel.

For Cobb the choice of Germany would not have been unexpected in the

1880s. Between 1860 and 1920 science played a major role in Germany, then leading the world in its system of higher education, especially in biological and medical research Middle class parents were sending their sons and daughters to acquire scientific training as a basis for their professional careers.

The choice of Jena would have had an added attraction for Nathan Cobb. It was the site of Carl Zeiss Optics factory and with the American's intense interest in microscopes and cameras it would have been a Mecca for him. His main objective was to pursue his studies with micro-organisms under the supervision of the outstanding figure in the field, Ernst Haeckel

Ernst Haeckel was Darwin's principal apologist in Germany. He championed an evolutionism that combined Darwinism, Lamarckism and idealist evolutionary theory. Beginning as a materialist Haeckel soon relegated natural selection to a secondary role favouring the Lamarckian principle of direct adaptation to the environment, a principle that more easily accorded with the ideas adapted from Goethe, the iconic German scientist and writer, that evolution was a progressive creative process.

The administration of the University was always keen to attract adventurous young scientists and by 1864 Haeckel was professor of zoology in a University which became a citadel of Darwinism where adventurers in botany, anatomy and physiology all contributed their critiques to the new ideas propounded by Darwin. Coinciding with the dissemination of the *Origin of the Species* in Germany, Max Schultze in 1860 introduced the protoplasmic theory of the cell from University of Jena. Schultze's theory gave Haeckel the starting point from which he developed his own theory that the granular substance of plasma was the equivalent of the first formed life: all organisms were plasmatic bodies differing only in degree of organisation.

Haeckel was also one of the first to describe progress in terms of cyclical growth. Haeckel's Darwinism-Lamarckian synthesis was translated into popular English textbooks and became immensely influential among biologists, psychologists and palaeontologists. He concluded that each modern embryo must pass through the embryological forms that represent past stages of evolution.

Haeckel had undoubted scientific gifts and he had a personality that helped to mould his biology into a secular surrogate for religion. He possessed a deep sensitivity for natural beauties, expressed in his scientific drawings, in his landscape paintings and his written accounts of his travelling. This passion for beauty in nature was combined with a passion for order in nature, which was reflected in his study of microscopic forms. He considered that 'wholes' or some 'wholes' are more important than the sum of the parts.

His star pupils, of whom Nathan Cobb must have been one, he called his *golden sons*. As a mirror of Haeckel, Nathan Cobb's work and opinions reflect this same love of order and appreciation of beauty in the nature of things.

Haeckel writings were not only popular but contributed significantly to what were then embryonic sciences of psychology, sociology and psychoanalysis.

Not everybody agreed with Haeckel. There was another man, August Weismann, at Jena 20 years later, who took issue with Darwin's position and questioned aspects of what could be called the 'Darwinian Creed'. He rejected the proposition that acquired characters could be inherited arguing that evolution took place by natural selection alone, operating at the microscopic level of the inherited germ plasma. Weismann stressed hereditary determinants while Haeckel insisted on environmental factors such as nutrition.

The University of Jena enjoyed a liberal, intellectual tradition having had Goethe as its driving influence. It is said that when Cobb submitted his application he also included some of his scientific drawings, which made a favourable impression on the renowned scientist. Cobb would have found an empathetic soul in Haeckel who was himself an artist of no mean talent. He later encouraged Cobb's art work portraying microscopic organisms, which would have accorded with his American student's own predilections.

By the time Cobb arrived in Jena in the mid 1880s, with his wife and three children, Haeckel had established his reputation as an original thinker on morphology and an expert on the microscopic life forms of the Protozoa. He had become a proponent of Darwinismus, which, for the German scientists, brought together a new philosophy of nature and new scientific theories about the protoplasm, the simplest form of life, *'the embryological theories of the recapitulation of the history of organic forms'.*

The Cobb family had sailed from America financed, not only by their own savings, but by a generous advance from their old friend Charles Prouty. In the time when they were living frugally in Jena, Alice Cobb copied passages from relevant texts which Nathan borrowed, not being able to afford to buy them. As he showed later in New South Wales he had the ability to grasp the material which came into his hands and to learn from other practitioners. He finished his studies within a year with a thesis about the nematodes found in whales, (in German), *Beitrage zur Anatomie und Ontogenie der Nematoden* and through Sir John Murray, a famous British oceanographer, impressed with his scientific art skills offered him a place at the British Association for the Advancement of Science Zoological Research Station in Naples.

In 1889, after working for the Station in Naples, he and his wife and three children, set sail Australia, considered by biological scientists one of the last frontiers of biology. Cobb disembarked in Adelaide, took the train to Melbourne and sought out Baron Von Mueller to present his letters of introduction. However, Von Mueller was not an enthusiast of Darwinismus and the fact that McAlpine had established himself in Melbourne in all probability, did not leave room for someone of Cobb's training.

He went on to Sydney. For a while after landing there was no work and not much money. He had come with introductions to business people as well as scientists, with the result that he found work with an importer selling Cashmere Bouquet soap for the American manufacturer, at the same time working on a series of illustrations for the publisher of *Picturesque Australia.* One can only speculate whether any of his drawings were reproduced in this publication because it is rarely that the artist's name can be found on the illustration.

In Sydney he repeated his product testing which he had done in Easthampton so that he could advertise soap. He was a salesman and market man who later found that the same techniques could be used to promote his ideas.

At the beginning of 1890 he became temporary Professor of Biology at the University of Sydney, at the same time being consulting vegetable pathologist for the new section of agriculture in the Department of Agriculture and Mines at salary of £100 a year. Finally, in August 1890, seven months after the establishment of the Department he became its Vegetable Pathologist, now combining investigation together with providing a consulting capacity for farmers. It was not until 1905 a short time before he returned to America that he found out why his permanent appointment was delayed so long. It appears the officers of the Department were a trifle wary of a man who had arrived with wife and three children with no plans in mind; so they sent to the United States to Easthampton to check his bona fides and be reassured that he was not a refugee from justice !

Head of William Farrer, sculpture by Rayner Hoff, 1937. For the Farrer Memorial Queanbeyan.

Wiiliam and Nina Farrer, Cuppacombalong about 1886

Courtesy Adrienne Bradley.

Wythmoor, Lambrigg Fell.

Courtesy, Janet Sutherland

Lambrigg Canberra, looking across the site of Farrer's wheat plots.

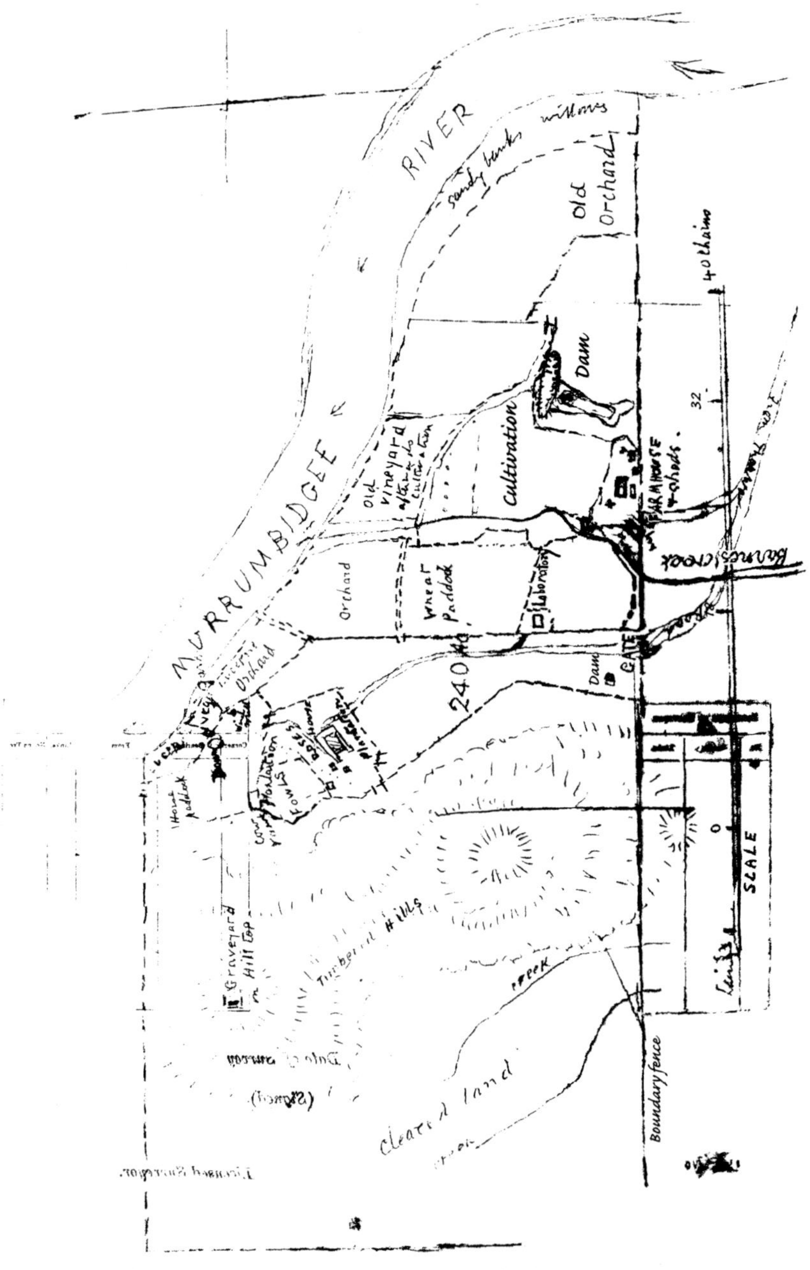

Sketch map made of Lambrig in Farrer's time, sent by Charlotte de Salis to George L. Sutton.
Courtesy, J S. Battye Library, Western Australia

Henri de Vilmorin
Courtesy, Sosthène de Vilmorin

F. B. Guthrie
Courtesy Mitchell Library

1897 Laboratory
Courtesy Riverina Archives

Guthrie's mill

Cobb in his laboratory at Bong Bong. He used the same arrngement at Wagga Experiment Farm.
artist E. M. Grosse.

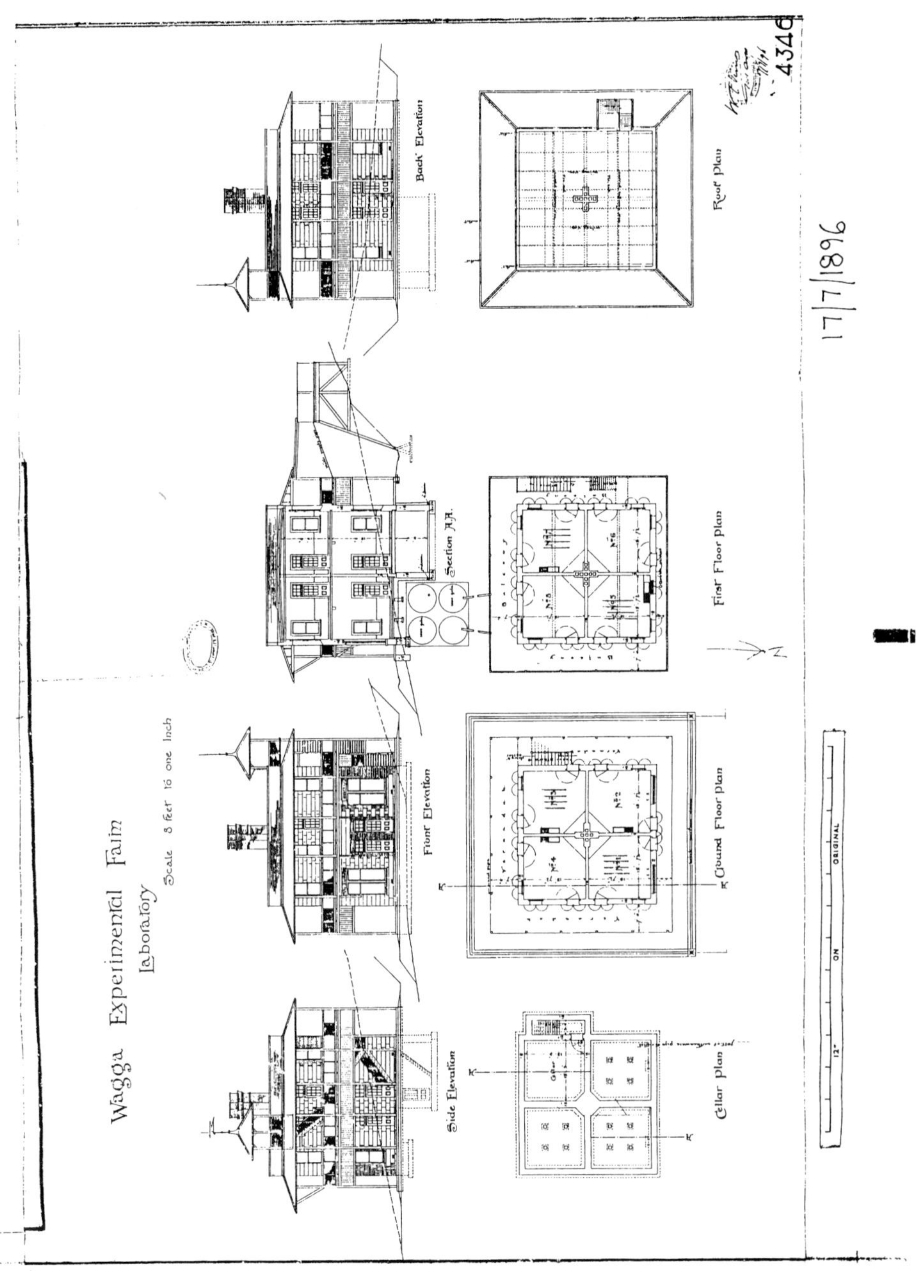

Walter Vernon's plans for the 1897 laboratory, Wagga Experiment Farm.
Courtesy, Department of Public Works.

1897: Wagga Experiment Farm. The granary at the bottom of the hill: the laboratory behind and the orchard buildings opposite

Courtesy, Riverina Archives, Charles Sturt University.

The experimental plots, established in 1893. Wagga Experiment Farm.

Courtesy, Riverina Archives, Charles Sturt University.

RUST 1890

THE PUSH FOR Federal government which Henry Parkes had outlined in the Federation speech at Tenterfield in October 1889, continuing together with strikes and bank crashes through the 1890s overshadowed a turning point in Australian agriculture which was taking place at the same time

All the concerns about the severe periodic bouts of rust infestation through the 19th century in Australia and other wheat producing countries came to a head in Australia in the disastrous season of 1889–90.

E.C. Large in his 1940 book 'Advance of the Fungi', wrote,

> *The greatest single undertaking in the history of Plant Pathology was to be the attack on the rust diseases of cereals. It was not to be done on paper or by laboratory research alone, but in practice, and over vast expanses of steppes, patunas and prairies.*
>
> *...The mighty rust investigation, soon to be world wide began in Australia with a series of Rust-in-wheat Conferences following the epidemic of 1889.*

The most fundamental information the 19th century on rusts was contained in Anton de Bary's book, *Morphology and Physiology of Fungi published* in 1866. In it he collated systematically, the knowledge which then existed. It was published as Part II of Hoffmeister's *Handbuch der Physiologischen Botanik*. De Bary, ' the young professor from Freiburg' became the leading authority on fungi.

Europeans and Americans knew that rust was hosted by the barberry bush over the winter in temperate climates but in Australia this did not happen, even on introduced barberry bushes. In India, the extent of the subcontinent provided varying climatic variations in the wheat growing areas. In the north the barberry bush thrived but in the south was not to be found.

In the 1890s botanists could distinguish two kinds of rust, *Puccinia rubegavera* on the leaf and *Puccinia graminis* on the stem, but the farmers could not. It was important to distinguish between the two–*Puccinia rubiga-vera* which did little harm in the early stages of the growth and the *Puccinia graminis* which fed on the nutrients in the plant which were then necessary for filling out the grain and the straw began to ripen. With out the plant nutrients at this point the result would be that the grain would be shrivelled and thin.

Confusion among the researchers occurred because the two rusts could not be distinguished either by sight or microscope when it appeared on oats, rye, barley or wheat and various grasses. The question which perplexed them all–were they all the same? Mark Carleton, as the result of his observations with the U.S. Department of Agriculture, carried out glasshouse comparisons and found that rust did not pass form one variety of cereals or grass with the exception in the case of wheat and barley. He concluded that there were differences between rusts–different varieties– and so he suggested in the face of dispute to call the wheat rust *Puccinia graminis tritici.*

Beyond this recognition of differences there was not much information to be obtained about rusts or what triggered a massive infection. As the letters to Australian newspapers showed every correspondent had a different theory which were reinforced by the anecdotal evidence which came to light in the first conference in 1890.

The 1889–90 rust infection of the wheat crops of the wheat-growing colonies caused an economic disaster and forced governments to take some sort of action. The Australasian Association for the Advancement of Science meeting in Melbourne proposed to the Premier that a Conference should be held on the matter. In South Australia influential men called on their Premier to take part in this initial conference in March 1890. As a result of this initiative eight delegates, two from each of Queensland, New South Wales, South Australia and Victoria, met in Melbourne. The convenors forgot Tasmania! For which the chairman apologised.

The proposal to hold the conference was greeted with certain reservations by the editor of the *Wagga Wagga Advertiser* of the day. Having commented that there were few topics which excited the interest of all the colonies outside politics,

> *The approaching 'Rust in Wheat' conference may be looked upon as a new departure in this direction, the subject being one affecting vitally what promises to become the largest interest we possess. It was at the instance of the Victorian Government the matter was taken in hand, and is gratifying to know that the wisdom of the step has commended itself to the various Australian ministers. Some exception has been taken to the appointment of one of the New south Wales representatives, of a gentleman whose experience in agriculture is not held in high esteem by practical men. It is certainly a matter of regret, that a mistake of this kind should have been made by the leading colony of the group. It is to be hoped that the other representatives will bring to bear upon the question, an amount of knowledge and experience sufficient to compensate for any lack of these qualities on the part of Mr Anderson*

The only report extant of the first meeting appears to have been published at the instigation of H.C.L. Anderson, the Director of the Department of Agriculture in New South Wales newly founded a month before the Conference. In the course of the discussions the experiences of the different colonies, as well as the United States and Britain were brought forward with suggestions such as the best sort of wheat to sow, when to sow and cutting wheat early. The consensus was that what worked in one place need not be right for another. Everybody seemed to recognise that there were many factors which were successful at one time or the other but what would really work was still out of reach. Professor Shelton[10] who had had experience in research in America outlined his observations and concluded:

> *...I want to join in what has been stated about the importance of facts in connection with this subject—one fact is worth 10000 theories, and we ought to have a systematic and very thorough set of experiments involving all these questions made under the supervision of men of scientific training, so that we may in reasonable time settle some of them definitely one way or the other.*

Anderson spoke after all the others and excused the department for not having made inquiries with regards to the problem as he would like to have done and said,

> *My own idea is that the greatest good we can do in this conference is to suggest some line of exhaustive experiments extending over at least five years, to be pursued in all the colonies simultaneously. Nothing is more confusing than to notice the results of experience conducted on the same wheat by different farmers on different soils. The climate and soil have an immense deal to with rust.*

Anderson's remarks highlighted the fact that wheat-growers were seeking for answers and the fact that the conference was convened demonstrated that it was necessary to forego inter-colonial rivalry and, as both Shelton and Anderson pointed out: find the facts as an outcome of scientific investigation. Anderson outlined some avenues of investigation especially the native source of the rust. The need for investigation on the matter lead him to promise the establishment of experiment farms in New South Wales. Although Hawkesbury Agricultural College was to be soon established, it was two years before the Wagga Experiment Farm was to be dedicated the first experiment farm, and then not primarily for wheat research but for horticulture.

Two letters were appended to the H. C. L. Anderson's Report on the Rust-in Wheat Conference. One was a letter from A. N. Pearson of the Victorian Department of Agriculture, written on the 5th May, addressed to J. L. Thompson the principal of Dookie Agricultural College in which he set out to refute what were Thompson's opposing views on the matter of rust:

> *You say you do not believe in contagion with regard to rust. But it has many times by very careful experiment been proved to be contagious. But no contagious disease is ever communicated unless the conditions are favourable to contagion...*
>
> *So in the case of the rust and wheat, the conditions for the successful devel-*

opment of rust, may when first considered, appear very simple. All that is necessary is the condition of the wheat plant favourable to the reception of the rust growth, and a condition of the rust favourable to its proper germination. Yet simple as these conditions appear, consider how very complex they are. For the wheat plant to be in a condition of receptiveness, it must be a suitable variety; then its soil, its manuring, its time and mode of sowing, its present stage of growth, the lowering of its general stamina hardiness and resistant power by present weather conditions, these and other circumstances must be suitably combined before it can be a proper condition of receptiveness.

… There is not, therefore the least difficulty in understanding the conflict of evidence given by practical wheat growers in regard to rust. To give a detailed explanation of all the apparently anomalous outbreaks of rust and of the equally anomalous escapes from rust would be an endless and not very profitable task. It is sufficient for practical purposes that we should know the real nature of the general problem, and that we should be acquainted with some simple and easily applied rules of practice, whereby the chances may be turned against rust. Such rules must not be expected to be infallible.

Thompson's perceived scepticism on the matter of combating rust has its reflection later on with Maurice McKeown when Farrer began his departmental work after 1898. Both Thompson and McKeown took the 'practical standpoint' rather than support the scientific experimental approach.

The other was Farrer's letter of the 7th May 1890. At this time he was still putting himself in the role of Blount's 'pupil' and Bancroft's assistant. He divided his letter into two parts, the first the argument for the way in which the actual quality of the seed and the second into the an outline of two preventatives of rust, again based on what he had learned from Blount and Bancroft.

Last winter I confined my sowing to between 30 and 40 varieties and gave special attention to the hybrids which have been created by my correspondent, Professor Blount, of Colorado, United States of America. Professor Blount, I may state, is the great authority of the United States cereals—a maker of new varieties by artificially crossing or hybridising and by selection… In this matter I have acted in conformity with Professor Blount's instructions.

He enumerated the qualities for which Blount was creating his wheats. They were all the attributes for which Farrer would strive: a grain rich in gluten, hardness, thinness of bran and shallow crease, right for the roller mills. Blount also bred for short stout stiff straw which was taken to give an indication of resistance to rust. In this first laying out of the desirable characters it was all directed to the reduction of rust damage to the crop. To the commercially minded farmers in the days of horse transport, hay was their most significant product of the wheat field but was not a prime concern to the wheat-breeder. Right through Farrer's breeding programme he was more concerned with the social value of his experiments to the community as a whole.

What emerges from this letter are the lengths to which the late 19th century Australian wheat breeders went in finding their way to the solution of the problems surrounding wheat cultivation. For instance, of three of Blount's hybrids trialled, one of them Farrer found of special interest as it was a cross made at his suggestion from one of Joseph Bancroft's rust proof Indian wheats with one of Blount's, (an inferior milling wheat with a superior milling sort), which after that was given the accolade by a milling expert. The exchanges which these men made of wheats from around the world and the work which progressed from them takes no account of time and the urgency of the need for solutions. The seasons north and south allowed for this sort of year long experimentation, disclosing to the searchers that what worked for one part of the globe might not proceed in another. according to the differences in climate and soil.

He concluded the first of his argument,

The above are my views in regard to the liability of wheats to rust and the grounds on which they have been formed. The immediate and exciting cause is, of course, certain conditions of environment—of moisture, temperature, location, and soil.

The conclusion of his letter was an invitation to farmers to try both 'Blount's Lambrigg' and Bancroft's Russian wheat, especially if *living on the eastern side of the coast range.* He went on,

If any gentleman resident in out coast districts has the inclination and the leisure for co-operating with me in my experiments, I shall be glad to correspond with him. Meanwhile, I am enjoying the privilege of receiving valuable co-operation from Dr. Bancroft of Brisbane.

By invoking both Blount and Bancroft he was lending authority and credibility to his argument. Dr Joseph Bancroft by this time had made his name in a number of avenues of medicine, business and investigations in agriculture.

By 1890 Farrer, among others, had no real solutions for preventing rust but he did have the basis for further experiments. In a sense he had been tinkering around the edges in a dilettante sort of way. While he was following Blount's practice, he was developing his own theories based on his observations.

It is interesting that Anderson included both these letters in the report he published on the conference—a contribution to the argument in favour of the Department's continuing to investigate the rust problem. The influential farmers and newspaper editors who read the report would have been furnished with evidence for continued work–and funding. After the Conference Anderson to moved quickly to have two key men made permanent members of staff, Nathan

Cobb almost at once and in 1892, F.B.Guthrie, both University trained in the sciences. Nathan Cobb as permanently appointed vegetable pathologist set about making his own investigations into the occurrence of rust.

Anderson in the course of the 1891 conference said,

> *My proposal is that the greatest good we can do in this Conference is to suggest some line of exhaustive experiments extending over at least five years, to be pursued in all colonies simultaneously. Nothing is more confusing than to notice the results of experiments conducted on the same wheat by different farmers on different soils. The climate and the soils have a great deal to do with rust.*

Anderson, as Director of the new Department of Agriculture, set about *eliciting all the definite information available with regard to the history of rust in this Colony* and collating it in preparation for the next Conference which was to be held the following year. The information was sought through the Agricultural Societies and a list was compiled of about 80 farmers who under took to conduct the Rust in Wheat experiments under the auspices of the agricultural societies, with seed supplied by the Department.

The Conference of Agricultural Societies of New South Wales held in the month following the Rust in Wheat Conference was used to promote the Director's planned approach. A member of the Agricultural Societies Council, F.B. Kyngdon gave an address to the meeting to lay the foundation for their co-operation. He dealt with the points which had been raised at the rust in Wheat conference in March: fertiliser trials; kinds and quality of the seed to be recommended; agronomic practices; value of burning stubble and straw after the harvest. He said that the subject of rust in wheat was one *deserving the closest attention of farmers and that in conducting such trials and tests the provincial associations could assist the Department of Agriculture.*

The first step for the Agricultural Societies was to distribute the questionnaire to the most active members, which related to the situation of the farm, the occurrence of rust at various times, weather details over the growing period and the farmer's agronomic practices.

A report had appeared in the *Sydney Mail* on the 14th June 1890, three months after the first Rust-in-wheat Conference listing the first appointments to the new department. The Director, H.C.L. Anderson on a salary of £800 a year, with R.L. Pudney, £350, and a temporary Statuistical Clerk, A.P.Reynolds, £200. There were two clerks W.Preedy and A.A.Dunnicliff; a Temporary clerk, G.Valder. Preedy was senior on £180 and Dunnicliff and Valder were each on £130. In addition there were two Professional staff at that time as consultants, Dr Helms and Dr Cobb. Helms was paid a fee of £150, while Cobb received £100. Both these men at the time were part time at the Univerity of Sydney.

To put in train farmers access to information, Anderson instituted the *Agricultural Gazette* drawing on the expertise of an increasing number of qualified people in the areas for which advice was needed

In 1891 Cobb had paid a visit to Cuppacombalong to inspect what William Farrer was doing. George de Salis in his diary recorded the visit:

2nd December,
Lea the entomologist in the garden collecting insects
3rd December
Farrer and Wills up tonight. Dr Cobb has discovered a new wheat disease
6th December Sunday
Dr Cobb up with all his microscopes and gave us an afternoon's entertainment
on the verandah.

Cobb in his report was much more laconic:

five weeks was spent by myself and Messrs Lea and Wills of this department
(who were appointed to assist me temporarily) at the farm of Mr Wm. Farrer of
Queanbeyan investigating wheat and its relation to rust. The results of this
investigation are in the process of publication.

The publication became *Contributions to the Economic Knowledge of Australian Rusts,* included over several instalments in the *Agricultural Gazette.*

F. B. GUTHRIE

I F FARRER WAS motherless, F. B. Guthrie had an over-abundance. His father married four times.

Where Cobb, Farrer, and Sutton did not have families with academic backgrounds, Guthrie's father was an academic at tertiary colleges during his working life.

Guthrie senior was born in London in 1833 and educated at the University School and College where his son was later to be educated. He became a distinguished mathematician, going to Germany at the age of 21 to study at Heidelberg and Kolb, taking his PhD before returning to London. In 1855 he graduated B.A. in London before becoming assistant to Professor Frankland in the Chemistry Department of Owen's College, Manchester

Four years later he was elected to the Royal Society of Edinburgh and assistant to Professor Lyon Playfair at Edinburgh University. After two years at Edinburgh he was appointed to a new post of Professor of Chemistry and Physics at the Royal College in Mauritius and in 1860 contracted a marriage with Agnes Bickell who was born in Germany. Professionally, he engaged himself in the new environment, sending papers to the Royal Society in 1864 and 1865 on *'Bubbles and Drops'*, as well as publishing a paper on *'Sugar cane and Cane sugar'*. Personally his life was not without incident. His son Frederick was born, according to the record in January 1862,even though Guthrie himself said he was born in December 1861. The January may be either the baptismal date or the registration date. Agnes Guthrie died in 1862 and Guthrie senior married again in December 1863. His second wife was the 18 year-old Laetitia Frost. They were only married two years before she died of tuberculosis. His third marriage to Amelia Smith took place in1865.

The family went to England in 1867 when Guthrie senior took some leave of absence from Royal College, but when in 1868 he resigned his post because of academic jealousies, they returned permanently to England. The following year he became lecturer and afterwards professor, at what was then called the South Kensington Normal School of Science, but later called the Royal College of Science. At forty, in 1873, he was elected Fellow of the Royal Society. This was a time of great interest in science by the general public and he founded the Physical Society of London. A measure of this growing interest in science and most probably the fact that Germany had become a leader in science education, in 1886 he gave three lectures on 'Science teaching' to the Society of Arts.

He became a widower for the third time. His marriage with Ada Amelia may have lasted longer than his two previous ones but there is no record of her death.

However in the census of 1881 he is included in the return for the household of William Buckingham as a visitor. Included also was another visitor, Eleanor Thompson who was single and 29. In the manner of romance novels one might conclude that she became his fourth wife. The name of his latest wife is not recorded in his biographical entry in the *Dictionary of Mauritian Biography*. When he died of throat cancer in 1886 his widow received a Civil List pension.

The boy Frederick was six when they went England and his life to that date may well have drawn him closer to his father, whose fondness for science, facility in French and German, and interest in writing poetry were communicated to the son. A succession of stepmothers would have strengthened this closeness.

F.B. followed his father to the University College and then in 1880, to a University in Germany, but not to Heidelberg. It was to study under Professor Zinke at Marburg preparing his doctoral chemical thesis. Professor Zincke had his own high teaching reputation and scholarly pedigree. He had been at the University of Bonn where he was an assistant to Professor Kekule who succeeded Justus von Liebeg famous among modern agricultural pioneers. Theodor Zincke was Director of the Chemistry Institute at the University of Marburg and among his students were not only Nobel prize winners in chemistry, but Nobel Prize winners in physics especially related to atomic physics. Marburg would have extra recommendation: Edward Frankland who had taken Guthrie senior as an assistant had been trained at Marburg.

However, Guthrie, the bright student, left after two years to take up the offer of Demonstrator in Chemistry at Queen's College, Cork. Professor Zincke told him that the university requirements could be met if he could initiate some research, supervised by the Professor at Cork. F.B. Guthrie never ever gave reasons why he left Marburg, nor was there any indication that he did indeed pursue his doctorate. One can only suppose that the reason he left Marburg was the reason many students leave their studies–money. Even at the end of his time with the Department of Agriculture the lack of money was acute.

His father's scholarly reputation may have enabled the son, in 1888, two years after his father's death, to be appointed demonstrator in Chemistry at the Royal College in Kensington under Professor Thorpe. He was in charge of practical teaching with Professor Frankland who had taken his father as his assistant in 1865. It was two years later in 1890 that, after being interviewed by a committee in London he was appointed demonstrator in Chemistry at the University of Sydney. He must have come highly recommended as Professor Thorpe was one of the Interviewing Committee. He took up his position when science was taken cross-faculty, Science, Engineering, Mining, Medicine and Arts. His arrival at the University coincided with the upheaval over the rust in wheat which was occupying both farmers and wheat-breeders. During 1890 and until 1892, the laboratory at the University was used by the Consultant Chemist, Dr Helms. When Anderson began to look for permanent scientific staff for Agriculture, Guthrie joined the Department in 1892.

F.B. Guthrie must share with Farrer the credit for the success of the wheat-breeding outcome of high protein, good flour colour allied to the drought resistance and high yield.

In 1891 Farrer's letter to the director, reported in the *Agricultural Gazette,* suggested that besides looking for rust-resisting wheat there should be some examination of the gluten content of the grain.

By 1894 Guthrie had begun the work with Farrer which involved the innovative method of testing the qualities of wheat which allowed Farrer to make educated judgements about the acceptance or rejection of the wheats he was using for breeding as well the new strains he was bringing into being.

It has been said that Guthrie stood in Farrer's shadow, even though Farrer acknowledged his debt to him, but he may never have felt that he himself was in the public's view standing in that shadow. In his own field he was highly regarded by his peers and was influential in the politics of his profession. In that day and age the profile of the man who bred the wheat which enhanced the economy and Australia's place in the world was well outlined only to those who knew about those things.

The years between 1892 and 1898 were marked by three men, Farrer, Cobb and Guthrie whose work became complementary. Each man's investigations were important in their own right but together the outcomes were to be important for everybody connected with the wheat industry; farmers, millers, manufacturers, consumers and the government. It was fitting that with the variety Federation, Australia should at the same time make an important contribution to world agriculture and economy.

From the time Farrer had begun to try to follow his ideas about wheat improvement he had written letters and exchanged seed. He wrote to Professor Lowrie at Roseworthy in July 1886, that he was preparing his ground work for selecting the most likely varieties:

> *I am distributing wheats as widely as possible as I can for trial in different parts of Australia and the reports I received I record carefully. I have already found out that the sorts that thrive best in Queensland are not the kind that do well here, also that the same wheats appear to suit the widely different climates of the coast and the interior of Queensland.*
>
> *...as my man failed to carry out my directions in planting them, the experimental plots will not be as reliable as I cold have wished them to be*

The work which he had done in the 1880s underlined the difficulties and the need to widen the number of qualities which would improve any new variety. Up to the time of Nathan Cobb joining the Department of Agriculture and becoming involved with the rust problem and the advent of Guthrie, he quite possibly no one with whom he could discuss ideas and the possible approaches to the solution. That there were three men to pool the findings enabled Farrer's progress with his experiments to be accelerated.

THE NEXT CONFERENCE

IN 1892 THE RUST-IN-WHEAT Conference was held in Adelaide, to which both Cobb and Farrer were invited as speakers. Following this conference the naming of wheats was given due importance with the election of a Nomenclature Committee, which was to be chaired by Nathan Cobb and on which William Farrer had a place.

The men who made up the Nomenclature Committee represented each of the Eastern States and came from differing disciplines. Cobb and Farrer represented New South Wales; an agricultural chemist, A.N. Pearson came from Victoria; Professor Shelton was an instructor in agriculture from Queensland, while South Australia and Tasmania were represented respectively by Professor Lowrie of Roseworthy Agricultural College and E. H. Thompson. It was decided to continue the experiment in each colony for one more year and present the results at the Rust-in-wheat Conference to be held in Brisbane in 1894.

Following the Conference almost immediate arrangements were made to put into operation the experimental plots in every colony. The list of wheats to be trialed were divided into categories. There two broad groups; those recommended for growing on a large scale and those for trialling on a small scale. Within the large scale group the varieties were divided into three further categories: rust-resistant; prolific and moderately rust resistant; rust into three further categories: rust-resistant; prolific and moderately rust resistant; rust escaping if planted early. The small scale plantings were divided into two, rust resistant and rust escaping. The seeds came from every colony, such as Blount's Lambrigg, Wards Prolific, Victorian Defiance, Queensland Defiance and Steinwedel.

In New South Wales ten farmers were selected to conduct a definite series of experiments: J.Faint, Kelly's Plains; T Bragg, Narromine; T Worboys, Orange; E Taylor, Young; H.D. Coker, Jindalee; J. Jones Senior, Berrigan; G. F. Berthoud, Corowa; W. Farrer, Queanbeyan. In addition J.T. Thompson, the Principal of Hawkesbury Agricultural College was also to conduct a similar series of experiments.

All through the 19th century the names of wheat varied from seed merchant to seed merchant, colony to colony and the delegates to the Rust-in Wheat Conference in 1892 had recognised that

The list of wheats to be trialed were divided into the categories Anderson had set put after 1891 There were two broad groups; those recommended for growing on a large scale and those for small areas. Within the large scale group the varieties were divided into three further categories: rust-resistant; prolific and moderately rust-resistant; rust-escaping if planted early. The small-scale planting was divided into two: rust-resistant and rust-escaping. The seeds from varieties such as Blount's Lambrigg, Wards Prolific, Victorian Defiance, Queensland Defiance and Steinwedel, came from every colony.

To carry out the experiments to achieve the best results, a set of rules was laid down, not only for New South Wales growers, but for all those taking part in the eastern wheat-growing colonies. The planting was to start on the same day in May in soil as nearly the same for each experimenter.

The rules laid out in the New South Wales *Agricultural Gazette* gave an incentive to the growers. If the wheat grower supplied the land, cultivated it under the inspection of, and according to, the advice of the Department he would receive a bonus of £10 (the equivalent of a workman's wages for about six weeks) and have the yield. The Department reserved the right to purchase any or all of the crop at a fixed price of 7 shillings a bushel (a capacity measure equivalent to about 36litres. Before sowing took place the land had to be in first class cultivation and the seed was to be drilled in, not broadcast. Every precaution had to be taken to keep the varieties separate, the weeds kept down, a diary kept and a report given to the department. The rules were uniform for every colony. The detailed instructions were sent to every participating grower and Mr Inspector de la Motte was detailed to co-ordinate all the administration.

The plots, the 546 varieties were planted on the same day in drills, each row 25 feet long, and 16 inches between the rows. Farrer's plots were the prototype.

While the farmers were participating, the research by Cobb and Farrer had other objectives,1)to determine how many distinct varieties there were; 2) to photograph each distinct variety; 3) to select from plants which were already well and favourably known those which could be improved in their desired qualities as well as their likely resistance to rust. In dealing with the first of these objectives, difficulty was experienced from the fact that in most cases the seed supplied was mixed or impure, Cobb saying a few years later that many diseases afflicting wheat came into Australia with the seed from other parts of the world. However both men and their assistants worked all through January and February to achieve their results.

GUTHRIE'S MILL

THE IDEA OF PRODUCING a nutritious loaf of bread was something Farrer, the one-time medical student, would have found to his taste. People of European decent in Australia at the time were bread-eaters and the quality of the flour from the wheat which was grown, always determined the flavour and quality of the loaf

In regard to wheat improvement there was a congruence of events in the time from1891 through 1892 and into 1893. Cobb visited Farrer at his experiment plots in 1891–92 and perhaps coming back off and on until after the harvest period, taking notice of what Farrer was doing. Not always agreeing with him and doubting that the process of experiment which he was carrying out might be a hit and miss affair. There can be no doubt that from the discussion of Farrer's philosophy as it related to his work he made it clear he sought to achieve: a product which would produce a palatable and nutritious loaf. Cobb's meticulous microscope work on the content of gluten and flour in a grain of wheat would have been in response to the questions for which the plant breeder was seeking a solution.

On a more practical level, Farrer had installed a bread oven at Lambrigg, in which he is said to have baked the bricks for the new house. Baking bread, combined with his observation, recording and analysing of the processes and the results pushed him to seek a finer, more precise way of assessing the quality of flour. As Leopold deSalis had earlier, Farrer would have bought his flour from Byrnes Steam Flour Mill in Queanbeyan. A bill was dated 1885 for the grinding of 200 bushels of flour @ 6 pence[11] a bushel. In 1884 there had been four consignments, February 4th, April 20th, July 10th and October 31st for 533 bushels altogether and in June there was a ton of second flour. This appears to be de Salis regular custom and so Farrer would have taken this flour for testing for gluten when he had the opportunity to consult Guthrie. Byrnes mill would have been grinding the wheat by millstones to a fine powder for customers like de Salis for family use and the second flour for the workmen and their families

Analytical chemistry had been practised in Australia before Guthrie arrived and had been used in Australia at Colonial Sugar Refinery Mills to test the quality of their sugar from the 1880s. Interested Australians like Farrer, privately and through libraries, had access to overseas scientific advances, so it was not a surprise that Guthrie should have been able to research information about flour and gluten. Guthrie was fully aware of the work done by the Danish chemist Johan Kjeldahl who in 1883 had developed a procedure which could be used for estimating the nitrogen content of foodstuffs.

Farrer wrote his letter to the Director about improving wheat for its gluten content in 1892: Guthrie analytical chemist, joined the department in 1892. That same year Farrer wrote to Vilmorin asking advice about a testing mill. Obviously interested, Guthrie set about finding a mill. Farrer now had another input into his search for the wheat with all the right characteristics for farmers, millers, bakers, and consumers.

In 1893 there was an item in the *Garden and Field* published in South Australia, relating to a sample of Blount's Lambrigg which had been milled in W.C. Harrison's mill, testing it for its flour. The report was unfavourable and contrary to the opinion expressed by an unnamed New South Wales miller which tested the same variety. A letter had been sent to Harrison's suggesting that the wheat tested *was not truly Lambrigg but some other variety.* Richard Marshall, a premier wheat breeder in South Australia, was asked for a sample of Lambrigg which he had been growing. Farrer concluded that the miller did not like the hard wheat.

When Guthrie joined the Department of Agriculture in 1892 he had a plethora of analyses to deal with: soils, manures, milk, wheat, sugar-beets, wines; water for irrigation and likely poisonous plants such as the Darling Pea. However he was soon involved in a project which helped to determine the direction wheat research would take. This was the devising of a testing mill.

In the annual report of 1897–8 he gave what might be considered a retrospective report on this development from 1894 when it was first mooted. Farrer, having had a relationship with the Department since 1890 had approached the Director seeking assistance with testing his certain of his wheats for gluten. Perhaps Guthrie had, had enough of answering farmers questions and the problem which Farrer presented attracted him.

The first samples were just ground to a meal, most probably in a mill that any farm, including. Farrer's would have had in the kitchen. Sifting the impurities from this meal would have enabled a reading of the gluten content of the flour to be tested. 1922 Guthrie recalled the beginning of the testing.

> *I was fortunate enough to obtain a couple of small rolls, such as were used for grinding small quantities of whole meal, and to succeed, with the cordial help of Mr R.W. Harris, head miller, Gillespie brothers in devising a method whereby the operations of a large mill could be fairly well imitated…Although with this mill we could not hope to obtain a flour of the high-class texture and bloom of miller's flour, we were nevertheless, able to determine with some degree of accuracy the important points of flour strength, gluten-content, and colour of the flour, as well as the proportion of bran, pollard and flour obtainable to bake the flour into loaves and to compare new varieties in this respect.*

However, it was rudimentary and it was decided that a more accurate and valuable test must be made on flour. The commercial mill would not consider taking the 12 ounce samples of flour for testing so the hunt was on to find a mill small enough which could produce from the small samples a flour as close as possible to that coming from the millers. Added to determining the gluten, other qualities could be assessed and importantly, the milling qualities of the different wheat varieties. The prospective milling quality was what would determine the price the miller would pay the farmer.

Guthrie set about making enquires and finally was able to acquire a pair of small mills from Austral-Otis Engineering Company in Melbourne. They would have served the same purpose for the machinery salesman as the cabinet maker's small-scale examples of his craft—demonstration of the product to the customer. These mills had come from Ganz and Company of Budapest, experienced and well-known makers of rolling mills in Europe since 1860. They had exhibited their machinery at an exhibition in Sydney some years before in the early 1880s

Farrer's wheats were both those locally grown and imported and by this time were some of his new crosses. Guthrie said later that Farrer's objectives were to improve these wheats to provide one or more varieties which could suit different soils and climatic conditions in the colony. Guthrie wrote in December 1901,

The study of wheat is one of the most fascinating subjects conceivable. Apart from the intense interest which necessarily attaches to anything connected with our principal foodstuff, the study of the development of the grain and of the operations necessary for its successful cultivation and subsequent treatment in mill and bakehouse, provides us with a series of problems of the highest intellectual interest.

…and all furnish material sufficient for a lifetime.

This appreciation of the stimulation associated with the investigations, ongoing and future, informed both Guthrie and Farrer in their co-operation.

By now Guthrie had interested two flour millers in his project, Brunton and the Gillespie brothers and they themselves would have had the new roller mills replacing the old grinding stones. To develop the concept which was in Guthrie's mind Gillespie's company lent him a miller, their mill manager, Richard Harris.

Guthrie was increasingly interested in the matter of strength of flour and its baking propensities, and, as his subsequent articles and papers showed, he set out to explain succinctly the reason for the investigation of the milled structure of the wheat seed as it related to the good nutrition of the community. By his work and the interest of the prominent millers, he drew attention to the objects of Farrer's research.

In his 1901 article which was based on a lecture he gave to the Royal Society of New South Wales, he described in simple terms the route of the grain through the mill. This was the process he had made it his business to learn at Gillespie's Flour Mill.

Before roller mills were introduced into Australia the grain was milled between two millstones which ground 'exceeding small', everything which was in the wheat sack, (weevils, grass and perhaps dead mice), to make the flour. The aim of the roller mill process was to produce from the grain a fine white flour, from which had been separated the bran and pollard as a marketable by- product of the process.

The first passing of the grain was through the corrugated rollers, set sufficiently wide apart to just crack and split open the grain. It was then passed successively through other corrugated rollers placed closer together. These were known as the 'breaks'. With this milling the bran was scraped off in flakes and removed by sifting through silk of sufficient mesh to allow the remaining semolina to pass through. This semolina was a coarse gritty white powder from which any particles of bran were removed by a current of air. The semolina was ground through a series of smooth rollers set closer and closer together. The flour from each roll was sifted through finer and finer mess until all the flour had been removed form the semolina, leaving pollard, part of the outer layers of the grain

The flour from all the different breaks was mixed together to make 'standard-grade' flour. The quality of the flour depended on the variety of wheat which was used.

When the test-mill was devised it was hand-operated making the process extremely time-consuming with the 13 steps with one set of rollers instead of millers multiple sets of rollers. To make it even more tedious, only about 12 ounces of grain could be milled at a time so when Farrer's samples were put through the mill, extreme care to save every last portion of flour. It would have taken too long a time for the impatient Farrer. The testing for gluten would have further delayed the results. The milling process was to prove to be a source of tension as time went on. To Guthrie it would seem that it was taking up too much of his assistants' time as the chemist's tasks were widening in scope. The timetable for the test milling was tight – between harvest and sowing Farrer would have wanted his new samples tested so that he could make his sowing and crossing plans.

Farrer thought that a full-time miller should be employed by the Department, suggesting that Max Kalhbaum be recruited for the job but it came to nothing.

The departmental 'millers' would have been subject a physical stress by the operation of the mill, which today would be subject of concern to the Occupational health and Safety Officer. The machine was fitted onto a wooden stand, bolted to a wooden floor resulting in noise and vibration underfoot.

By 1898 the manual process was replaced by a Pelton engine driven by water and in May that year Farrer wrote to Max Kahlbaum, *Mr Guthrie wrote me last week that his new mill could be ready by the end of the week.* The mill proved itself, not only Farrer but other farmers, buyers and bakers found its results financially rewarding for the them.

After twelve years the mill was supposed to be fitted with a gas engine, but whether this was done is not recorded.

By 1912 Guthrie in writing *Wheat and Flour Investigations* detailed the progress in the acceptance by people actively involved in the wheat and flour industry had accepted the whole concept of investigation. However, the department itself seemed less involved with wheat as a major area of investigation. Perhaps it was felt that the School of Agriculture within the University of Sydney was the proper place for investigation. Pridham was no longer designated wheat experimentalist but plant breeder. Although he continued with new varieties of wheat, he turned his attention to new oat varieties.

THE NAMING OF WHEAT

ONE OF THE MAIN purposes of the trials at Lambrigg was to begin the differentiation of the names of wheats. Farrer more than anyone else had imported seed from experimenters overseas and they would prove to be a benchmark for the task. Some of the wheats arriving in Australia in the 19th century travelled an interesting path towards their part in improving grain.

William Farrer from the time he became interested in the problem he saw in wheats obtained from France, India and the United States, in return sending wheats developed in Australia abroad for trial– back to these same places and also to places such as Kenya even though it was not a wheat-growing area. In his collection he had 300 samples.

The firm of Vilmorin-Andrieux in France had published a catalogue to which Farrer referred for his orders, but he also sought the advice of Henri de Vilmorin on suitable varieties for his purposes. When he sent some of his own varieties to de Vilmorin, reports came back on the trials of Farrer's cross-bred wheats. An example of this exchange is the variety, Cailloux. This wheat was a cross between Farrer's wheat Florence and Aurore made by Professor E.Schribaux in 1920. Aurore was selected in France by Vilmorin in his own experiment plots, from one of the unfixed samples sent by Farrer to Vilmorin. The unnamed variety was a cross between Jacinth(one of Blount's wheats) and Ladoga which had come from Russia. In 1922 this seed was sent to Tunis, and Cailloux was selected from a crop by a farmer of that name. It was introduced into Australia by the New South Wales Department of Agriculture in 1932 and was then grown on the North-West Slopes of New South Wales where it gave good results when compared with other varieties being grown in the same period.

Farrer's first introduction to Indian wheats had come through Dr Joseph Bancroft, a founding member of the Queensland Acclimatisation Society. Bancroft, assisting the Queensland Government to find a suitable wheat for the climate, had been sent Indian wheats by a friend, Dr Dymock who was in Bombay. Until Bancroft's death he and Farrer continued the work on wheat, adding varieties which he obtained from America.

Farrer's main source of American wheats was from Professor A. K Blount of the Colorado State Agricultural College. A variety called Allora Spring was an exception, but had a connection. Allora Spring, which Nathan Cobb said on one occasion had come from California, had an interesting scenario. William Farrer in a letter, written on the 17th February 1896, to W.C.Grasby of Roseworthy Agricultural College in South Australia, remarked, *This variety is the very first wheat I introduced into Australia.*

The name Allora was taken from the wheat growing district on the eastern Darling Downs, which leads one to suppose that Farrer introduced it for Bancroft in the mid-1880s before Farrer himself was fully experimenting at Lambrigg. In all probability it came from E.W. Hilgard, at time establishing an Agricultural College and Experiment Station in California, and having connections with Blount.

Like many other wheat varieties of the day this one had many names for the same grain. In the case of Allora Spring, one of its synonyms in California was Sonora. As Sonora it had another connection with Farrer; he introduced into Australia a Blount variety called Amethyst which had Sonora in its breeding–Lost Nation x Sonora–and then used in his breeding programmes.

The variety Thew which Farrer named in 1890 combined in its breeding wheats from France, United States and India. One parent was a cross between Blount's Lambrigg and a Fife wheat and Sinew of Farrer's own making. Sinew itself was across between a French variety, Blé Carré, and Hornblende, one of Blount's wheats. The other parent was a cross between Fife and Hussar, a variety of Farrer's bred by crossing Hornblende and Indian G. Thew was an example of the wide scope of the varieties used by Farrer in striving to combine those most likely to achieve his desired outcomes over a number of seasons.

Farrer and Bancroft were not the only ones to import wheat to find those most suitable for Australian conditions. Often in 19th century Australia new varieties of plants were grown in botanical and private gardens. In South Australia Dr. Schbomburghk was the director of the Adelaide Botanical Gardens where he established an Economic garden for growing plants with an economic value and suitable to grow in South Australia. In 1881 he imported from South Africa the old wheat variety, Du Toit. He sent small samples to interested farmers, one of whom was James Ward. The plot which he had planted In his garden was rusted except for one plant, the seed from which he planted the next year. This wheat became Ward's Prolific, the ancestor of later varieties. The other imported variety was early Baart, another old South African variety said to have been grown in south Africa from the mid-18th century. It was introduced by Professor Custance, the first principal of Roseworthy Agricultural College in 1884. These South Australian imported wheat strains became part of the parentage of wheat varieties developed by Richard Marshall at his farm at Wesleys in South Australia. Marshall worked contemporaneously with Farrer, the two meeting at the Rust-in-Wheat Conference held in Adelaide in 1892. Their friendship lasted for many years.

All through the 19th century the names of varieties wheat varied from colony to colony, seed merchant to seed merchant and person to person. It was often mixed before the farmer bought it. Dr Bancroft told of getting some wheat seed from a handful of birdseed. Up to the end of 1890 the varietal nomenclature of wheat was haphazard and uncertain, listed in the catalogues according to the common name given to the wheat by various seedsmen, and farmers in different parts of the colonies. There was no definitive catalogue of wheats such as

Vilmorin's. A differentiation between varieties was based largely on the appearance of the plant and the grain. The delegates to the Rust-in Wheat Conference in 1892 recognised that some attempt should be made to rationalise the varieties which were under over 600 names in Australia. The decision was made that this should be done in each state. All the wheats were to be planted on the same day, the 21st May, 1892. Every Colony was to test for the same characteristics under similar conditions: *milling value, rust-resistant qualities and periods of ripening.*

In New South Wales, Dr Nathan Cobb as chairman of the Committee was to supervise the operation when William Farrer offered Lambrigg to be the site for that colony's trials. While Farrer had the cultivation and general care of the plots, Cobb was delegated by the Conference to make a careful examination of the potential rust resistance of each plant. And to investigate the cause or combination of causes which might enhance his character in the wheat. Farrer imported named varieties provided a benchmark for the rationalisation.

When the definitive list came to be drawn up it was found that the nearly 600 varieties they had started with could be brought back to 300.Instead of the faulty recognition criteria used before, size of seed, beard or beardless, growing habit, colour of leaves, the arrangement of the seeds in the ear and height of the plant were all part of the description.

So that the conditions in each colony would be uniform, the plots were of equal size in approximately similar kinds of conditions. 546 varieties gathered for all wheat-growing colonies were planted on the same day in drills each 25 feet long. There was just one problem as far as the object of the research was concerned–the season was good and very little rust appeared. That did not stop the research work of Cobb and Farrer. They had other objectives, 1) to determine how many distinct varieties there were; 2) to photograph each distinct variety; 3) to select from plants which were already well and favourably known those which could be improved in their desired qualities as well as their likely resistance to rust. In dealing with the first of these objectives, difficulty was experienced from the fact that in most cases the seed supplied was mixed or impure, Cobb saying a few years later that many diseases afflicting wheat came into Australia with the seed from other parts of the world. However, both men and their assistants worked all through the summer months of January and February of 1892–93 to achieve their results.

Good milling quality was number one among the qualities they were looking for and any other attributes were put under the heading, *other striking qualities*. The wheats which attracted attention were, Blount's Lambrigg, Smith' Nonpareil, Steinwedel, Rattling Jack and others derived form Purple Straw.

After the Lambrigg experiments, samples of 375 wheat plants were brought back to Sydney, to the Department of Agriculture, with ten stools to each sample. The purpose of this was to supply each colony's Department of Agriculture with samples of wheat in exchange for theirs. Beside the wheat plant samples, the heads of the varieties were also bagged so that seed could be distributed to farmers who wished to establish private nomenclature plots, or learn to distinguish the different whets by name.

The naming of some varieties became a point of contention between Farrer and Cobb as evidenced by the letter Farrer sent to Cobb about Defiance and Blount's wheat– *I have never received the faintest suggestion from anyone except Campbell that Blount's Lambrigg was identical with Defiance.* Macindoe in *Notes on wheat breeding and varieties in Australia,* said of Defiance that it originated in the USA as the result of a cross between Golden Drop and White Hamburg and that it was introduced into New South Wales in the early 1880s and was probably one of the parents of 'Blount's Lambrigg'. This is where Farrer disputed Cobb's naming.

When Farrer took issue with Cobb over the identification of 'Defiance' in his letter to Cobb on the matter he said that, *Defiance had been selected in 1878–79 before 'Blount's Lambrigg' was named.* The history of the Colorado State College in the course of outlining Blount's work, said, *Supposedly, this grain was originated by a man named E.C.Pringle from whom Blount obtained a small sample in 1879.* Blount worked on developing a wheat which by 1885 had a head which could hold an average of 43 kernels as against 21 of the original sample. In Colorado 'Defiance' was considered Blount's most notable contribution.

Cobb by a series of tests which he devised together with observation and consultation, reduced the number of varietal names from about 600 'named' varieties to approximately 300. After over seven years work and perhaps attendance at the Centennial Nomenclature Conference held in Paris in 1900, he published his results between 1901 and 1905.

When the definitive list came to be drawn up it was found that the nearly 600 varieties they had started with could be brought back to 300.Instead of the faulty recognition criteria used before, size of seed, beard or beardless, growing habit, colour of leaves, the arrangement of the seeds in the ear and height of the

Nathan Cobb always appreciated devices and modifications to existing apparatus so as to facilitate operations. He took some ideas from Farrer on the matter of setting up the experimental plots such as the surveyor's chain for measuring out the rows: a dibbler with the depth of the hole for the seed measured so that only went in to that depth and the bottom of the dibbler the right width for the hole. For himself he devised his own adjustments to cameras and microscopes to made into pieces of apparatus to deal with out of the ordinary aspects of his work.

He owned all his own equipment and chemicals which he used for his work. In the Department's first years neither he, nor anybody else for that matter appeared to have any equipment supplied. It was only about 1896 97 that the Department began to spend money on laboratories and equipment. In his scheme of education microscopes were a necessity for which he pressed to be provided for the students at the Wagga Experiment Farm. Although he modified his equipment when he needed to it was never makeshift.

In 1893 he set about making 2000 tests for measuring the hardness of wheat. It had always been the practice of the miller to do the bite test on the grains of wheat to make his judgement about the grain. If the grain was hard to bite it was hard to mill. It was time consuming and obviously depended on the state of the miller's teeth. Cobb decided that he could devise a tool which would make the test more accurate. This is how he described it in his departmental report:

In consequence, he took a pair of cutting pincers, filed off the edge until they were as about as blunt as the front tooth of a middle-aged man, and with these pincers, so altered, proceeded to bite the grain of different varieties of wheat, the force which was required to bite the grains in two being measured on an ordinary spring balance.

He said the results of his tests were 'promising'. He said that they accorded with 'ordinary opinions' on the milling qualities of wheat and Corresponded 'very fairly, with the chemical tests made by Guthrie. In addition he maintained that they would be time-saving and cost-saving, because chemical analysis was time-consuming and more expensive because of the cost of the chemicals 'and so forth'.

Cobb's reports of what he had done often lead to the conclusion that he felt he needed to justify not only his time but his greater value to the department.

While Farrer was doing his wheat crossing and Cobb casting a doubt over the prospect of valuable varieties of wheat emerging in the foreseeable future: 'by selection from the …crosses…some valuable wheats may be produced', Cobb set off on a path which he had devised. He began to concentrate on the seed. Each variety and each season produced seeds which differed in their appearance. Because there was controversy about which size, shape and plumpness of seed was the best to sow, Cobb made some sieves which would grade the seed he had st the Wagga Experiment Farm. He described what he did:

The wheat sown this season has been graded by means of hand sieves made by Dr Cobb for the purpose from what is known as 'half-round brass wire'. The messes of the sieve are elongated, their width being graduated most carefully from 3_ millimetres down to 2 millimetres, the different grades being known as 325, 300, 275, 250, 225, and 200, meaning 3.25 millimetres, 3 millimetres, 2.75 millimetres, 22.50 millimetres, 2.25 millimetres and 2 millimetres. These six sieves give seven grades of wheat.

Having shaken his wheats through the sieves he set out a table of calculations for the 27 wheats he graded, giving the total amount in ounces of each variety he put through: 14.96 ounces of Allora Spring had some seeds in every category except 325 and even one score under the heading of 'Rubbish' The varieties which scored the highest in 325 grade were Algerian(18.89 0unces graded) and Grosse's Prolific(17.23 ounces graded) with 5.76 ounces each.

Like Farrer discarding his crosses which did not go through Guthrie's mill with satisfactory results, Cobb used his grading system to discard wheat which did not score well. He was not entirely satisfied: the sieves themselves he made initially were not proving accurate enough when he wanted to do further work on the grading of wheat, when some doubt was cast upon the original work with the sieves, his results being disregarded to a large extent by both farmers and millers. He was so convinced of the value of sieves for those experimenting with wheat that he proceeded to have more accurate versions made. In 1897 he reported,

> *I might add that during the last five years no instruments in the Wagga farm have been in such constant use as these sieves, and it is with great pleasure that I can now say that similar sieves are being used elsewhere in consequence of the publication of the results of the Wagga Farm experiments.*

Cobb working on the experimental plots at Wagga Experiment Farm reduced the number of names of about 600 'named' varieties to approximately 300 after over seven years work. He published his results between 1901 and 1905.

It may have been this work with the sieves which prompted one of his party pieces which was recorded by one of his American colleagues, and apparently worked up in his early years of return to the US. Edna Buhrer a later colleague described one performance

> *Dr Cobb arrived at the specified hotel at the last moment in a taxi and came into the hotel carrying a suitcase – some say up to four suitcases under the weight of which he staggered in, refusing help from the taxi driver, doorman or any of his fellow diners. When Cobb's turn came to contribute to the programme, he cleared a big space on the table, put the suitcase on it. He then explained that he had a hobby of which few people knew and as far as he knew nobody else had this hobby. he collected holes! As part of his collection a prized Australian fenc-post hole: prized because there were not many fences in Australia. However he had a fossil woodpecker hole – one of the gems of my collection. It dates way back into geologic ages – almost as when first there was any wood to peck. It took months of careful labour to remove the useless stone from around that hole so as to leave nothing but a carefully prepared hole. But it was worth it. I've given this hole a room to itself… you should see how many people come to see that room. They come in flocks to rave over that hole. In bated breath of course.*

This sense of humour has all the ring of an Australian origin and one of the leg-pulls familiar to the readers of the *Bulletin* and the bush yarn-spinners. It was suggested that he used this performance to satirise scientist and non-scientists who often talked about things thet knew nothing about.

NINA DE SALIS AND FARRER

THE LAND WHICH was to be called 'Lambrigg' had been bought by Leopold de Salis at an auction of Crown land held in Queanbeyan, for £240–£1 an acre. It was along the Murrumbidgee River from Cuppacombalong and taken up as a conditional purchase in 1876 for his unmarried son Rudolph who died later that year. On the 6th February 1892 Leopold de Salis transferred the title to his daughter, Henrietta Sarah Nina Sophia Farrer. This was at a time when the financial restraints on the family may have persuaded Leopold de Salis to secure a place to live for his daughter.

Henrietta Sophia Nina Charlotte was the only daughter of Leopold de Salis, the son of an aristocratic Swiss family which had for many years lived in England. Her father had come to Australia in the late 1840s followed by his elder brother William who had financial interests in banking in the colony. Leopold took up land first at Junee and then at Darbalara near Gundagai, in partnership with a man called Jack Smythe. While he was at Darbalara he married Charlotte Macdonald, the daughter of another station owner.

His brother William wrote about Leopold's wife in 1845, *We hear, however, that she is very economical and keeps a bright lookout after the house, sheep and farm*

In 1846 Leopold wrote to his mother, *Our first years here and when everything was going against us and I saw nothing before me but a dreary struggle...I determine to content myself by marrying and I did so, thank God. I have now a very sensible wife on whom I doat(sic) every day more... And since than the times have bettered and Charlotte has bettered the management, and she who denied herself everything to keep me out of debt, now that we can afford it is insisting on what in England is termed comfort and decency, but here luxury.*

In 1847 when they had dissolved the partnership Smythe returned to England and deSalis ran his properties with the assistance of his brother-in-law Charles. In1846 Leopold had made a decision to return to England when he had acquired enough money so in 1855 he sold Darbalara and Junee proposing the leave when the transfers were completed. While waiting for everything to be propitious for the little family to make the voyage, he bought Cuppacombalong near Queanbeyan as a place for his family to live. In the meantime his mother had died. He saw no reason then to leave Australia.

Having settled at Cuppacombalong he began, as did many others, to acquire land under the Act Regulating the Alienation of Crown Land which was passed in 1861, using conditional purchase and selection often in the name of his workers on the station. Receipts for the regular payments appear in his papers in the National Library. It was during this time that 200 acres was obtained for his son Rudolph on the western side of the Murrumbidgee, the future Lambrigg.

These properties on the Limestone Plains were not the only interests which he had. He invested in property at Bowen and other districts in Queensland which carried in all over 80,000 cattle and 20,000 sheep and later, after Nina and Farrer were married and he was away on his surveying, she travelled with her father on visits of inspection in 1883–84. Later deSalis was involved with his sons and Pierce Smith in property in North Queensland.

Leopold deSalis was not only a man of property but he began a Parliamentary career in 1864 when he was elected to the Legislative Assembly in the New South Wales Parliament in 1864 where he remained until 1874 when he was appointed to the Legislative Council.

When Charlotte deSalis, Nina's niece, wrote to George Sutton to give him some background for an article he was writing on her aunt, she was not quite sure when she and William Farrer met. She thought it was at Duntroon. She did say that they had both been engaged before. The death of Nina's mother, Charlotte, in 1878, and Nina's promise to her father that she would never leave him, may have put any thought of marriage aside. She was now the mistress of Cuppacombalong with all the responsibilities which went with the role in which she would have been trained by her mother. From the accounts of her by others later, she also inherited from her mother a firmness of purpose. Farrer right up to the day they were married always called her Miss deSalis, with the respect due to the 'lady of the manor'. If he was talking or writing about to anybody else he always referred to 'your sister' or 'your daughter' and later as the Missus.

Nina de Salis, although an adept at the household management delighted in the out-of-doors; she rode and played tennis and was known to go away for several days on camping trips. She and her brothers were keen photographers, doing all their own processing and one of the gifts which Farrer gave her when he married her was a new camera. The photographs which are in the National Library collection demonstrate and eye for composition and an ability to record a less serious approach to events.

In February 1882 Farrer was staying at Cuppacombalong as a letter to *The Australasian* testifies, and later that year he and Nina were married at St Phillip's in Sydney. It was not an elaborate ceremony, only members of the family attended. He was later to write in a letter to Mark Carleton that his marriage was the greatest good fortune which had happened to him.

From the time William Farrer and Nina de Salis were married in 1884 and while he was still active as a surveyor, he began his horticultural activities on Lambrigg.

Before the House at Lambrigg was built on a slope looking south-east to the hills beyond Tharwa, the land was made up of several 40 acre homestead blocks. The land was bisected by Barnes Creek from which the clay was later dug for the new house. South of Barnes Creek was the old farmhouse and all its outhouses

Between the farmhouse and the Murrumbidgee River was a dam and beyond that the original orchard running along the sandy beach of the river. A cultivation paddock, where Farrer grew rye for the horses was in front of the farmhouse and was bordered by the creek Originally, before the Farrers lived on Lambrigg, there was what Charlotte de Salis labelled on her sketch map as the 'old vineyard' which lay between the cultivation and the river. This southern part of farmhouse, sheds, dam and vineyard and orchard seemed to form a separate section of the property. paddock off the south. According to Pridham the laboratory which was built in 1900 overlooking the wheat at its western end close to the road into the farm.

It was, however six years before the Farrers began to build on the property and only a short time before Cuppacombalong had to be sold. The new buildings and improvements from 1892–93 were north of the creek on land rising gently towards timbered hills. When the house was built, and before the deodars grew up to obscure the view, Farrer could look out and see his wheat experimental plots.

Behind the house the land rose to a sheltering hill recalling for Farrer his birthplace in Westmoreland, Lambrigg Fell. All the land between the house and the river was given over to orchard, vegetable garden and lucerne, with water conveniently located in addition to the river, by a dam and a windmill. Nina Farrer, with the help of a farm worker, looked after the vegetable garden where the seeds sent from overseas were planted, cultivated and enjoyed by the extended family. It was a self-sufficient enterprise which provided a have for a man who for twelve years was subject to hostility and ridicule outside its gates.

A tree planting programme was planned and an outer windbreak of *Pinus insignis* [13] was planted behind and to the south of the house as a windbreak against the cold winds off the mountains. In the early 1900s, within this line, deodars were planted. Farrer grew them from seed sent as a gift from Mr Moreland a plant-breeder from India, who had come to study Farrer's methods in 1900. Cypresses and almond trees from the Wagga Experiment Farm were planted, but none excluded the view across the wheat plots.

From the time before Nina had married Farrer, he had bought and propagated roses to plant at Cuppacombalong and so it was at Lambrigg: rose beds were planted along the north side of the house, many of them from Cuppacombalong. In front of the house shrubs were planted and like the trees did not obstruct the line of sight to the wheat plots.

The house appears to have been built in two stages. The lower storey was completed at the end of 1892 when the Cobb party arrived from Sydney and the upper storey was finished the following year. At Cuppacombalong Nina and William Farrer had lived in one wing of the house while Nina's brother George and his family lived in the main part of the house.

Leopold de Salis first return to England was in 1893 with Nina and his son George. When they returned in 1894, it was to move in the new house at

Lambrigg. With Cuppacombalong sold, George and Mary de Salis and their children moved to Lambrigg together Henry deSalis and his family.

Later, in 1900, according to Pridham, a laboratory and storehouse was built nearer the main house with *3 rooms pisé. 6' verandah. Gal. iron roof. Well built stone foundations, 13' walls. Stone floors. Spouting round the verandah. Loft over two back rooms.* Charlotte de Salis, in a letter to George Sutton said she could remember seeing a hand operated winnowing machine standing on the verandah.

Nina kept a well-ordered house at Lambrigg, as she had no doubt had done at Cuppacombalong with strict meal times. Her niece, Vera, corresponding with George Sutton when he was writing an article on Nina Farrer, recalled breakfast time at the Farrers. Farrer had already been at work in his laboratory since about six o'clock but returned to the house for 8 o'clock breakfast. Not only was Farrer a stickler for punctuality but Nina liked everybody in the house to be on time.

We used to sit down at the dining room table. My aunt sat down at one end. She had the proper things laid out. The copper kettle, on its stand with a light underneath it, the teapot on the other side. There was always a cruet of boiled eggs on the table. It was a rather nice looking metal cruet, which held about six eggs and was always full. There was always plenty of toast and butter. She had porridge. She liked the coarsest oatmeal porridge she could get.

You could tell she wasn't very pleased if you were late for breakfast, but she never said anything. I used to have an egg and toast. I suppose I was hungry and I used to think 'Oh, I'd give anything for another egg'.. But I couldn't build up courage to ask.

In spite of what would have been straightened financial circumstances, Nina maintained the household standards she had known at Cuppacombalong, obviously enough to intimidate her nieces. Her care for her husband's diet and health was profound. For visitors the hospitality to be found at Lambrigg was legendary.

The house at Lambrigg after World War I had to yield to the exigencies of the Commonwealth Government valuation with regard to Nina Farrer's continued occupation. The objective description by the valuer in 1920 said:

The main house: Well built brick house, 13'walls. Gal iron roof,. Stone foundations. 12 rooms (brick). Upstairs 5 rooms(Pise). 8' walls in front in basement. Upstairs rooms plastered walls and metal ceilings except library and kitchen. One side of the library of wood. 9' verandahs and 2 W.Cs. principally of wine cellars.

VISITORS AT LAMBRIGG

IN THE SUMMER OF 1892–93 Nathan Cobb and his staff went again to assist with the harvest of specimen wheats, test and take photographs. Farrer drove to the platform at Tuggeranong to meet the mail-train, expecting to meet Dr.Cobb, but returned alone. Cobb and brown, his assistant, and arrived about midady and picked up a lift with one of the locals who took them down to Lambrigg, across the low summer Murrumbidgee. The next day Farrer and Cobb drove the wagonette back to the train stop to pick up Mrs Cobb and the four children including the baby along with Miss Massie, her companion. While the children stayed at Cuppacombalong homestead, Mrs Cobb was taken over to the original farmhouse at the southern end of Lambrigg.

As well as the cobb family, a journalist from *The Sydney Mail* was witht eh aprty. The time had arrived for the publicity to be put in train on what was happening in the matter of the country-wide wheat experiments concerning rust. Nathan Cobb was proving himself an accomplished publicist.

Farrer's trials were of a scientific nature; selected, meticulously observed and recorded. He was carrying them out on the lines he had basically laid down in his letter to the 1890 Rust-in-Wheat Conference. However, since that time he had demonstrated his technique to Nathan Cobb in 1891–2 and in discussion with him perhaps finding himself making modifications in what he was doing. Farrer's was not the only site for trials. In 1892 a number of selected farmers had been sent directions which went out over Anderson's name but had bo doubt been devised by Cobb and Farrer. Seed was distributed and results tabulated and later compared.

According to Lane's report of the effort at Lambrigg, there was in New South Wales a three phase process of scientific examination in the selected areas, involving the millers and 500 farmers. The Farrer trials marked the scientific phase with the close examination and analysis; the farmers 'participation was the complementary phase and the millers processed the resulting rain. The outcome of the planting was to have each varieties of wheat identified under a single name, instead of two, three or four different names, according to the colony, district or even the farmer.

George de Salis described in his diary how Cobb, with the assistance of Grosse and Lane took photographs of each kind of wheat from two directions as well as of sections of grain put under the microscope, the photographs formed part of the brief from the Nomenclature Committee in seeking to rationalise the seedsmen's lists. Sketches of the work in progress were reproduced in *The Sydney Mail* together with photographs of some of the most promising wheats. All this work in the field was carried out under canvas sheltering all the apparatus they were using.

Farrer's experimental work was done totally at Lambrigg before any seeds were taken for Nathan Cobb or farmers to try. He used the old farmhouse as a makeshift laboratory and storehouse. *That was the room Mr Farrer had put up for his own use when he had a caretaker-cook in the house, so that he could sleep there if weather-bound,* is how Mrs Cobb later assigned it. It was described rather prosaically in the valuation report on the property in 1920 as having, *6 rooms pisé. Gal.iron roof, stone floors, no ceiling, spouting on the south side 6' to 9' walls. 2 brick chimneys.* Because the conditions of selection said that the land must be occupied, it was home at one time by a succession of married couples It was here that the Cobb family stayed in the summer of 1892–93.

Mrs Cobb left a description of her time at Lambrigg when the family lived in the old Farmhouse or *shack* as she called it. Like most recollections written down as they tell it the story goes backwards and forwards.

In December 1892, N. was doing wheat experiment work on the farm of Mr William Farrer, in or near Queanbeyan. That was before an experiment farm had been started by the new Department of Agriculture. His farm had been cut out of the ranch belonging to his wife's family, the de Salis family.

Pleasantly situated on the farm, near the experimental work, there was a simple plain building, a shack perhaps, put up as a residence for a caretaker and his wife. At that time it stood empty and Mr Farrer the use of the building to N. if he would like to have the family come up to occupy it. The idea of getting into the country in hot weather appealed to me strongly, and preparations were started at once. 'My lady help' was to go, for with a six-weeks old baby, and five other children, the eldest of the six a little over eight and a half, it was too big a proposition for me to undertake alone.

… At the Farrer farm all food would have to brought in, so I made quite an extensive memorandum of all the non-perishable foods I could think of. This was mostly dried and canned goods, cheese, jams, biscuits–or crackers. The canned goods were mostly fruit, for with the exception of tomatoes, no canned vegetables were to be had.

…The ranch house was five miles distant. The family sent over a cow also a mare and buckboard for our use. My lady help, Miss Massie was of Tasmanian birth and accustomed to country conditions in that part of the world, and was a good driver, so we had use for the mare and buckboard. Our drives were through 'the bush',–very much so. I remember driving where there was only a path, and in some way the front wheel was one side of a sizeable log and the hind wheel the other side. There had to be some backing done, but Miss M. with no concern apparently, considering it quite a joke, although some of the children, including the baby, were along. But the steed was steady, and Miss M. was careful when steering our way between standing trees. There were no dense woods in that part of the country.

Mr Farrer had men working for him, and one was detailed to look after the cow and do the milking. He also collected the eggs, which was some job, for the hens with so little human companionship were wild, seldom coming near the house. The children would have liked to hunt for eggs, but that would not have been allowed on account of the danger from snakes, the bite being poisonous.

An assistant of N.'s, Mr Lane, father of a family of nine, was a great help about getting breakfast, while we were finding kitchen help. When I asked him to make the cereal thicker, he said that he was accustomed to making it thin at home, to make it go round. Afterwards it was thicker. Mr Lane and another assistant wee camping, getting their own meals, but of course when Mr Lane got our breakfast he was invited to share it and I think he enjoyed it.

Nina Farrer was known for her care of those who needed it and there would be no doubt that it was she who would have organised for Celia Tong to come to help, as Mrs Cobb recalls,

After a few days we found Celia, whose home was within walking distance, who was willing to come and do the general work and cooking. Miss Massie took care of the children with my help, and did the bedroom work. I have no recollection of the laundry work, but as our water supple was limited, and I do not remember any hardship in connection with getting the work done, there must have been some family within reach where it could be done.

In spite of help, accommodation was cramped and would have been regarded as pioneering. In hindsight recounting it to her daughter, there may have been a smile recalling how the family had managed.

The shack, the building in which we lived for six weeks had a very large room which we used for living room and dining room, one large bedroom which accommodated Miss Massie and the four children; a small room opening out of the large room had been built on at a later date. It was large enough to hold a large size double bed, a bureau and a chair. I took charge of that room which was occupied by my husband, myself and the baby. There was an outside door to the room which was fortunate for the room was so tiny I had to make the front side of the bed, go around through the living-dining room to the outside door, go around that end of the house and enter the little sleeping room in order to make that side of the bed...

The kitchen was a large room with doors opening at the front and back and had a cement floor. These three large rooms opened on to a floorless porch, the roof of which gave the three rooms some shade form the hot sun. The large bedroom—also with a cement floor contained a single cot for Miss Massie, another very large bedstead which was occupied by Margaret and Frieda with Ruth between them, all sleeping crosswise on the bed.

Mrs Cobb, although originally a country girl, may have forgotten how necessary remedies could be conjured up away from town:

To make a bed for the two boys, two carpenters horses were brought in, boards laid across them and a mattress tick filled with fresh straw placed on the boards. All three of these rooms were unfinished—the timber all showing. Opening out of the kitchen on the shady side o the house was a store room where canned goods and anything in receptacles could be kept, but except for bread, butter, etc we lived in a 'hand to mouth' manner, food being prepared for a single meal, or for only one day at a time. We seldom kept cooked food overnight.

In the kitchen there was a colonial oven and a camp oven. The colonial oven was a square iron box setting back in the fire place and bricked in at each end. The main part was the oven; underneath was for the fire and on top also a fire was made which helped to heat the oven and could also be used to heat a saucepan, frying pan, or any open receptacle. Some iron rods rested on the bricks at the side, which were built up a little above the top of the oven and on this a receptacle could be placed for boiling, frying or broiling could be done. The meals were ordinarily cooked by the colonial oven.

The camp oven, placed over a fire in cement or bricks was like a huge cast iron kettle with three legs long enough so that a fire could be built underneath. A rack at the bottom of this oven separated the loaves of bread, each loaf in a separate bread tin. A flat cover could be put over the top of the oven and meat kettles were boiled on top of it.

In spite of 'pioneering' conditions everything seems to have been done to ease Mrs Cobb's life with her very young baby,

Miss Massie, my lady help, who went with me from Sydney, was born and brought up in Tasmania, I suspect under many pioneer conditions. If Celia needed extra help in the kitchen Miss Massie supplied it. I have seen the colonial oven in operation but I am not sure that I ever saw the camp oven in use.

Nathan Cobb and his wife had common interests and even under the primitive 'holiday' living co-operated in preserving the memories of the stay:

We decided we would like to have a souvenir of our pleasant sojourn at Farrer farm, but neither of us had much leisure time, so we selected a view and N photographed it, as an aid to my memory, and I took notes as to colours, and when we were back home I started a small oil painting. With N. as critic, I produced a picture which both were glad to have.

The time spent at Lambrigg was not all work on the tasks of the wheat plots. George deSalis proved a considerate host, and along with Nina, who was still living at Cuppacombalong, organised excursions and entertainment for the Cobb family as well as Cobb's assistants.

Cobb himself provided entertainment in reverse. A keen photographer, George deSalis recorded in his diary how Cobb took *instantaneous photographs* of everybody in the garden when the family came up for the afternoon. George and Nina were used to the difficulties of the old style of camera where the subject had to remain as fixed as a statue while the photograph was being taken. These new photos with their spontaneity amazed them.

A STICKLER FOR JUSTICE

W HEN FARRER RETIRED to Cuppacombalong in 1886 he became involved in the activities of the other run-holders on the Limestone Plain as well as his own researches. One wonders at his physical condition. The photograph of him taken in about 1886 shows him tall and thin and appearing pale. George Sutton commented on the diet of sugarless food and careful preparation by Nina Farrer suggesting that the hard living conditions of his surveying days had left him physically depleted. His friends and associates later commented on the care with which Nina attended to his well-being aided by the vegetable and orchard which she maintained and the poultry she kept.

One of the controversies which occupied the gentry at that time was Frederick Campbell's putting up a fence and closing a road which by custom had been used by anyone passing along that part of the plain as a right-of-way. Everybody was drawn in from top to bottom. Samuel Shumack gave his account of the attempt by Thomas Southwell to knock out the fence posts. Frederick Campbell subsequently brought a case against him of assault. In the climate of the time Campbell succeeded and Southwell was made bankrupt. All sympathy lay with Southwell, one of the farmers in a small way.

In the de Salis collection there is a series of letters between Leopold deSalis and Andrew Cunningham on the controversy; included is part of a letter from Farrer on the matter in very much the same style as the letters he was to write to the *Sydney Mail* after the 1890 Rust-in wheat Conference.

Charlotte deSalis recalled two occasions when she was a child when she Farrer responded to what he saw as injustice: once when he had an altercation with a man on the Queanbeyan train and continued it on the platform, punching him to the ground and the other was an explosive exhibition of anger at the dinner table at Lambrigg. This occasion was the newspaper report of the conviction in 1894of Alfred Dreyfus, the Jewish French army captain for allegedly passing information to the Germans. He was dismissed and deported to the penal settlement on Devil's Island off the coast of Guyana in South America This became the cause célèbre of the day because it was considered by many to be a vindictive political pursuit of an innocent man because he was a Jew. When Dreyfus was sent to Devil's Island many worked hard for his release and his reinstatement. The fight for right, justice and truth would have appealed to Farrer as a man to whom perceived injustice was an anathema.

Farrer saw further evidence of what he would have considered as injustice in years during which his wife's family were entangled with property and legal problems

Such were his feelings about the law and justice that he wrote a *Letter to the "Justice" Committee Federal Convention 1897*, now held in the National Archives.

> *I venture to bring before you the following suggestions which I think deserve consideration of your committee. The first is that in cases where laws enacted or adopted by the Parliament of the Commonwealth are in any way obscure, imperfect or out of harmony with other laws in force in the country, the expense of the 'test' cases should be borne by the commonwealth. Also in order that the Commonwealth may have to bear the cost of a few such 'test' cases as possible and that any laws enacted by the Parliament may be in the least possible degree in danger of being obscure, imperfect and out of harmony with other laws in force, it appears to be highly expedient that any law has received the sanction of both Houses of Parliament and should be submitted to a standing committee of Judges, eminent jurists before it becomes law.*
>
> *In regard to the 'test cases', I venture to submit to your consideration an opinion that it is at least an important to a country, that an honest suitor should not be ruined or even suffer on account of any obscurity, defect on incongruity there may be in a law (for the which the Parliament not the suitor is responsible) as it is that a prisoner should be committed for a crime, when the evidence against him leaves any doubt of his guilt.*
>
> *…live by the laws and are interested in their being uncertain and defective affords sufficient reason for the propriety or rather necessity for this provision.*
>
> *In the event of your committee having completed its labours and sent in its report, I beg you will hand this letter to the President of the Convention*
>
>

I am Sir,

Faithfully yours,

William Farrer

Sometime between 1899 and 1901 he had occasion to write a memo to the Under-Secretary of the Department of Agriculture in defence of Maurice McKeown, because he felt that justice had not been served. There is no date, but it is possible that it was written in perhaps in 1899–1900. There is a letter from McKeown in the letter book in which he angrily queries a movement against him by Mr Preedy who was at that time Chief Clerk. However, whether this addition to the papers relates to Preedy's charges is not known.

> *Copy To be attached to papers 65/3409*
>
> *It is very right and just to Mr McKeown that I should attach to the above papers, in which I have stated that the veracity of my late assistant Mr Carl Franzen was absolute and I consequently indignantly repulsed Mr McKeown's implication, but this was not the case, a statement that his leaving me was the result of my discovering that he had been guilty of a series of untruths. The opinion that I hold is that up to the time of taking the first false step, his truthful-*

ness to myself had been absolute and that the telling of the first untruth led to his getting into a deep mire of additional untruths. Probably, the necessity I had impressed upon him when he told me of not having been absolutely truthful was the cause of his making the unwise and…reckless efforts he did to conceal the lie. This regarding Mr Franzen I much rather have left unwritten but the obligation to be just to Mr McKeown is paramount and it is for this reason alone that I have penned this memo.

FARMERS AND EXPERIMENT

THE GOVERNMENT experiment farms were not altogether regarded highly, Stewart Mowle, a Member of Parliament, in a letter later in the 1890s was still regarding them then as fads. However with the energies of Cobb and the making of silage, orchards and stock quality, they were proving themselves to farmers.

Sutton, writing in 1913, when he had become Wheat Commissioner in Western Australia wrote,

Scientific agricultural work of this kind was not then looked upon with the same respect that it is today…In the mind of th average man scientific agriculture was not worthy of serious consideration. The sterling value of Farrer's own work has been largely responsible for a changed attitude of the popular mind

Farmers were market conscious and looked for wheat of high productivity. By the 1890s wheat productivity in the Australian colonies had declined. The increasing productivity in the last decade of the 19th century and the first quarter of the 20th century is considered to be due to 4 factors:

1 The introduction of new wheat varieties
2 The more extensive use of fallowing
3 The application of fertilisers
4 The introduction of mixed farming.

These were to remain to be implemented following a resolution at the 1892 Rust-in- wheat Conference to rationalise the names, under the supervision of a Nomenclature Committee under the chairmanship of Nathan Cobb.

While the farmers, by and large had one aim in view–productivity value at the point of sale, Farrer had a more philosophic text to follow over and above the production of a rust resistant wheat–to provide a more nutritious and appetising loaf. However, as his quest for rust resistance concentrated on the characters of the plant he came to see that this objective might exclude what could be the real value of the resulting grain. He was developing varieties which had potential but it was only when he could co-opt the assistance of F.B. Guthrie that he could take the milling quality, the amount of the gluten in the grain and exclude any which did not demonstrate these characters.

Farrer had imported wheat himself and had been given seed by others who had done the same thing. In his collection he had over 300 varieties which he had labelled with the names under which they had been supplied. When the definitive list came to be prepared it was found that throughout the wheat growing colonies over 600 named varieties had to be sorted. Even among ones which looked on first glance to be the same, minute differences helped to distinguish them. Size of seed, beard or beardless, growing habit, colour of leaves, arrangement of the seeds in the ear, height of the plant, were all part of the description.

Nathan Cobb had set about the task of finding how to distinguish between varieties and catalogue them under group names. In addition, what his work set out to achieve was the possibility of predicting from a small sample of grain what the commercial value would be. This prediction could be achieved under his plan without running the field and milling tests.

Cobb made comparisons of the internal structure of the seeds of many of the commercial varieties, using his microscope techniques and painstaking dissection of the seeds. The differentiation between seeds lay in the aleurone[14] layer where there was a variation in the ratio between cell cavity and cell wall. Some cells were small and others were large while the cell walls could be thick or thin. Examining the cells of those wheats where analysis had shown them to be rich in protein, the aleurone had large cells with thin walls and the gluten was not in the aleurone layer but in the protoplasm of the flour. Cobb had brought to bear his experience under Haeckel. Cobb was able to go from this finding to predict from any sample from any variety the strength of any flour which would be produced.

In order to make an estimation of the comparative amounts of cell wall and cell cavity in any particular grain he used the slide preparation technique which he had perfected. He took a slice of the seed, mounted it in carbolic acid to clear it and with a magnification of 100 diameters, he photographed it. The negative was projected onto a 75 centimetre square piece 'stout' tissue paper laid over a glass screen with a magnification of 1500 diameters. The next step was to race the outlines of the cells and cell cavities onto the paper.

Frieda Cobb Blanchard in her biography of her father described how she and her siblings were involved in this work. She said she could remember the big glass screen which was on a large closed-in balcony of their home in Sydney but she recalled the next step more clearly:

With a scratch or retouching tool (like a pencil with the lead replaced by a sharpened steel) the inner cell wall lines were cut, releasing the 'cell cavity walls'(55% to 70% of the paper area) and leaving a network of paper 'cell walls'; and the weights of the two then were compared. The cutting was our job.

It was an after school job when she and her three sisters sat at the round dining table with its smooth wooden top and each with an aleurone layer, a soup plate and cutting tool. The breeze was shut out and as each time of work was up, the papers in progress and those which were completed were, with the most extreme care were carried to the empty guest room and laid out on the beds.

Nathan Cobb acknowledged his daughters painstaking work when he published Universal Nomenclature of Wheat:

This method gave greater speed, ensured greater accuracy, and gave a chance to introduce lower-priced labor, and is the best I have so far been able to devise for the purpose.

His labor costs were threepence a sheet– the price of a loaf of bread.

WAGGA EXPERIMENT FARM

THE ESTABLISHMENT OF the Wagga Experiment Farm as a model farm– a demonstration farm– was not the result of any ideological interest in promoting farmers and settlers best practice on the part of the Government of New South Wales. Ministerial responsibility for agriculture was forced upon it as the result of political pressure applied by rural interests beyond the pastoralists who were the majority in the Legislative Council.

Both Victoria and South Australia had had Departments of Agriculture and Agricultural Colleges which served as research centres from the mid-1880s. While the New South Wales Government was being made aware of the fact that it was lagging behind these other two states in grain production and other agricultural pursuits like viticulture, the Legislative Council was dominated by the men of property, in this case the pastoralists, wary of the radical nature of the would-be agriculturalists. The selection process had lead to farming being carried on as far west as people could get land; land beyond the coast and cooler slopes. Newspapers, in both city and country, pressed for agricultural education for both boys leaving school and men already farming. Economically, the State needed to feed itself instead of relying on grain from the other colonies or even from America.

New South Wales agriculture was in the doldrums, maintained by outmoded practices, crops and land use. On the other hand, South Australia and Victoria were moving on with new varieties of wheat, improvements in farming practice and mechanised farming. In 1884 when Victorian farmers moved into the South-western Riverina they came with their machinery as well as their household goods. The Riverina was becoming the farming heartland of the colony, with Wagga Wagga the growing town.

Even when new machinery and fertilisers had been introduced adequate means of communicating knowledge had not been put to work to educate farmers to the new crops and methods. The lack of will to redress the situation was evident in the lament of the New South Wales Agricultural Society Council in 1872, when it concluded that there was little prospect of doing anything for agriculture in the shape of technical education. Selectors Associations and Farmers Unions began to exercise pressure for scientific agriculture to be promoted. The settlers with few financial resources had eagerly taken up land under the various Land Acts from 1861, being able to support themselves and their families working on stations and keeping a few livestock, with very little application to farming. However it has been estimated that by the end of 1886 that the work of fencing, building and other improvements for the pastoralists, like tank-sinking had been completed. The settlers now found that their main source of income would have to be farming, about which they knew little.

Cultivation of wheat which had begun in the first days of the settlement in the Sydney basin had moved to the highlands and slopes and at last found the most suitable climate and soil on the western slopes of the Great Dividing Range. Nevertheless, even in the most favourable of places the grain crops were subject to the various diseases and pests and in addition farmers could not maintain their crop yields with the practice of continuous cropping. There was for them no advice offered to improve their practice.

Newspapers too, both city and country brought pressure to bear on the government to provide technical education and practical advice for farmers. The only Technical College in the State was in Sydney, although by the end of then 1880s travelling experts visited places like Wagga Wagga to give lectures on such topics as chemistry in agriculture.

Ministerial responsibility for Agriculture was given to the Minister for Mines, making it a combined portfolio. This was the position in early 1890 when the new Department came into being with H.C.L. Anderson as the Director. His appointment was criticised the press because he was not considered practical enough: he was an arts graduate from Sydney University with a major in agricultural chemistry. A month later he was to lead New South Wales delegation to the Rust-in-wheat Conference in Melbourne.

A report appeared in the *Sydney Mail* of the 14th June 1890, three months after the first Rust-in-Wheat Conference in Melbourne. The fact that Mr Pudney *'had reported on several sites for experimental farms which owners had agreed to sell to the Government'* was no doubt intended to assure the farming community and provincial newspapers that some moves were being made to take some steps in the matter of the rust problem. He had, however to accede that he had not yet made any decisions. As it transpired the first experiment 'farm' was to be Hawkesbury Agricultural College. The first experiment farm to be established specifically was at Wagga Wagga on a resumed temporary common and its first plantings were not for wheat experiments but for an orchard.

Under the new Director it was determined that there should be a central agricultural college with satellite Model Farms. Hawkesbury Agricultural College was established in 1891 and as many as 20 sites were inspected around the State for model farms, which were considered to be suitable for their particular district.

The farm which was to be the first in the State was called the Murrumbidgee Experimental Farm and intended as a Model Farm. Although the first report to appear in the *Agricultural Gazette* was headed Murrumbidgee Experimental Farm, it was subsequently and thereafter Wagga Experiment Farm.

The site of the Farm, on Crown land in the Parish of North Wagga, including part of the North Wagga common, had been decided on at the end of 1891, but the legalities took until late in 1892 to complete. It was dedicated at last in

October 1892. The reporter from *The Burrangong Chronicle* in Young would write that the in its natural state was no better, if as good as the neighbouring districts and in the days of selection had been passed over.

The address of the Farm for many years was Bomen. When the site was selected, the Hampden Bridge had not been built and the town of Wagga Wagga, five miles away(about 10 kms) was reached by a low-level toll bridge about 100 metres from what is now the new Wiradjuri Bridge. An alternative to this crossing if you were disinclined to pay the toll was to ford the river near the Black Swan Hotel in North Wagga. The original plan for the Farm was to for it to be horticultural to demonstrate the cultivation of fruit and vegetables. Outside a few stations with their Chinese gardener, vegetables were not grown and scurvy was a persistent problem among the rural families. Vegetables such as melons, peas, potatoes and turnips were tried for their suitability in the soil and climate so that they had the possibility of being a quick answer to the needs of the farming community. Fruit trees were a longer term project for any farmer as well as for the Farm.

The Farm covered 2000 acres and the most prominent features of the landscape were the Two Sisters hills. The boundary on the southern side is marked today by Farrer Road, across the hill to the Pine Gully Road. The eastern boundary was the travelling stock reserve on the then non-existent Coolamon Road. A. C. Benson, the fruit expert who came to the Farm in 1894 to set up the orchard said that the persistent weed problem in the orchard came from it having been a sheep and cattle camp.

It was a virgin site. There were no buildings, no fences, no water. John Coleman had been appointed the first Superintendent to initiate the project which was to be the model for the establishment of the subsequent farms. He began by having the eastern side of the hill cleared. This was done with difficulty as it was all green timber and none had been ringbarked. Initially, he cleared 75 acres for the orchard and 10 acres for Dr Nathan Cobb for his experimental wheat plots following the commitment by Anderson to resolution of the problems with wheat. The suitability for soil and climate were the same for the wheat varieties as for fruit and vegetables and Cobb set about trialling all the wheats to which he had access. With exception of the regimented setting out of the vegetable and wheat plots the rest of the farm was operating on an ad hoc basis. While the Superintendent lived at the homestead 'Estella' about two kilometres from the Farm, other officers made-do in tents and bark huts. Any of the workmen employed lived in North Wagga.

The first dam on the Farm was constructed on the little creek which flowed at the bottom of the cleared sloping ground, at the end of Cobb's experiment plots, now Booranga Estate Vineyard. Later, on what Coleman described as the farm proper, along the Pine Gully Road, land considered most likely to be successful was cleared for both stud wheat and trial crops of grain.

When Cobb began his experimental plots with the aim of rationalising the various wheat varieties available for farmers, he reserved on acre for experimental planting of Farrer's crossbred to estimate their suitablity to the soil and climate of the Wagga Wagga district.

George Valder who arrived in 1894 to be the manager of the orchard, later recalled that he wondered at first what the Department was doing when it sent him and A. C. Benson there. The plant available to them for work was a team of 11 yoked oxen, two draught horses and a farm dray besides a plough and a harrow. When it rained they slept under trees and when they wanted water they sent down to old Mr Matthews who had an adjoining forty acre block on the travelling stock route. They moved to a cow shed and the cows got under too. When he asked the Department to build him a fence they allowed him £14,which would have allowed him to employ four workmen for a week. The proviso was that he sold firewood to cover the amount.

As the orchard was being laid out, Cobb had arrived to make preparations to harvest his grain and found that there was no storage space. He reported that he had requested a grain storage shed, 20feet by40 feet. It was put up with 'great celerity' a month after ministerial approval had been given and painted and finished throughout so that it was decided to hold under its roof the first field day held on the Farm. This was the large shed in which Valder was able to live and work in after John Coleman left in 1895 to go to establish Bathurst Farm.

THE FIRST FIELD DAY

FROM 1889–90 AND the failure of the wheat harvest, the following years did not favour an upturn in the value of wheat as a crop. In 1893 John Coleman the Superintendent of the Wagga Experimental farm gave a lecture in Wagga Wagga, sponsored by the Murrumbidgee Pastoral and Agricultural Association, with the title, *Crops other than wheat which can be grown profitably in the District.* Wheat which a decade before seemed the most profitable and simplest crop had lost its attraction. From the agricultural point of view other crops of monetary value and fertility-producing value needed to be cultivated to reverse the declining commercial cycle. Wagga Wagga had for many years valued the agriculturalists settled round the town, because the family units provided a stability which would have been lacking otherwise.

Coleman opened his lecture,

To begin with wheat is the main staple crop grown in this district and the question to ask ourselves is does it pay better than other crops, and is there a probability of its still paying.

He quoted *Coghlan's Statistical Register for 1892* which forecast that there would be a need to import 250,000 to 800,000 bushels before the next harvest, and also showed a decrease in the import figures from 1890 and 1891. In 1890 the country imported two to three million bushels of wheat and in 1891 from one million to three million bushels. The forecasted harvest of 1893 would result in over-production with the resulting fall in prices. Over a decade the acreage in the Murrumbidgee district had increased from 4375 acres to 93,449 acres but without commensurate increase in the market.

The fall in the price of wheat with the oncoming harvest was further exacerbated by the decline in the quality of the grain because of poor farming practices. It was found that farmers were not changing the varieties they were sowing nor were they sowing the pure seed which made for mature and heavy crops. The seeds which were made up of mixed varieties resulted in crops of uneven heights with undergrown crops ripening at different times. The farmers were putting in the same time, effort and money but not improving the profitability.

He highlighted the experimental work being carried out by Nathan Cobb with 400 or more varieties planted at the farm, pointing out that what they were doing was not farming but experimenting. In saying this he was trying to put at rest farmers concerns, later echoed by McKeown concerning Farrer's experiments, that experimenting served no educational purpose for farmers as they did not demonstrate profitable methods.

Coleman went on to promote crop rotation for soil rejuvenation and as practical alternative sources of farm income,

> *I would grow the crops somewhat after this way, wheat, peas, malting barley or oats; vetches or peas to plough in, summer fallow, ploughed about four times, the wheat, then over again. Now I am quite convinced this could be carried out and the land at the end of ten years would be as good or better state than it would be at the start.*

For his audience he elaborated on the marketable value of alternative crops, such as peas, oats and barley, citing the profits to be made from these by not relying on imports from the other colonies or New Zealand. It seemed obvious that he was opening up new venues for the farmers about which they were ignorant. He continued elaborate on the value of growing potatoes and vines *You will always find a market if the produce is of sufficient quality.*

The newspaper report recounted the discussion on experiences of farmers planting alternative crops, leading Coleman to comment that inviting a discussion he had set the ball rolling.

While Coleman could lay before his audience his ideas, it was not until 1894 that there could be any sort of practical demonstration on the first field day at the Wagga Experimental Farm of the proof of his words.

The field day held at the Wagga Experiment Farm was held on Saturday December 4th 1894, a little over a year after the first experimental plots were laid out. The day was not a propitious choice. Being Saturday farmers would normally be in town, their time devoted to business and sales. However, advertisements, press releases and *Wagga Wagga Advertiser* editorial a week before, all indicated that it was to be a day of importance to farmers and not to be missed.

In March and April of 1894 the newspaper editor hab been critical of progress of the Farm's, taking the Department to task, although he did concede that the Department had 'exerted itself' to obtain seed varieties from different countries but at the same time regarding it as premature to think of success in finding a solution. He went on:

> *But it may be said that strong hopes are entertained that the series of experiments which are being conducted by Dr. Cobb the pathologist, will eventually result in giving to the farmers a variety of seed that will prove invulnerable to the disease.*

Then a warning sounded:

> *The low prices which have ruled for wheat during the past harvest have naturally had the effect of bringing home to farmers the necessity of turning their attention to some more profitable method of utilising their land and labor.*

John Coleman later in the year had a meeting in Wagga Wagga under the auspices of the Murrumbidgee Show Society to talk on this very topic. In April the paper returned again to the matter of advice and guidance for farmers:

It is now four years since the Department of Agriculture…decided that the vicinity of Wagga was the most suitable centre for the establishment of an institution of this kind for this portion of the colony.

He echoed a sense of frustration which seemed to exist among those men who thought that Wagga Experiment farm may not have been becoming the centre of value to the farming community that was promised.

These farms were in the first place, meant, by means of experiments carried out by practical experts, to demonstrate to agriculturalists the crops best suited for the different districts, and secondly, it was intended they should afford facilities for the theoretical and practical instruction of young men desirous of devoting themselves to the business of farming on scientific lines

The editor acknowledged that John Colemen had done a good job with the funds available to him but,

…all the experimental work of value yet performed has been in the direction of growing and testing a number of varieties of wheat on a strip of ground cleared for that purpose and although we admit the value of the efforts made to discover a rust-resisting species of this cereal, it cannot be contended that the other purposes of the farm could not have been proceeded with at the same time. We are now told that steps are shortly to be taken to lay out a vineyard and orchard, but the wonder will be why both grape-vines and fruit trees were not planted at lest two years ago. The fact is, the project has been starved from the first.

Although the editor might take issue with the lagging department, H. C. L. Anderson was to say later that the Department had not been formed under an Act of Parliament and in consequence its purposes had not been defined, and it was not until it was separated from the Department of Mines in 1908 and Anderson had been re-appointed Director that its direction and structure solidified.

There is no doubt that the first field day was in large measure a response to such critical comment and the spur which saw the seed shed erected with such 'celerity'. Such a place for a presentation instead of a bark lean-to or a tent would have made some correction to the poor opinion which may have been present in the community. Nathan Cobb knew how publicity and demonstration could help sell an idea as well as a product, being a persuasive tool in winning acceptance. He had done it in the past in Easthampton with testing every kind of product for local manufacturers in his home laboratory. He had done it in Sydney when he had to sell Cashmere Bouquet soap. Now he knew he had to show the practical men, and those who thought along those lines, that experimental processes were not a waste of time and money. The rust problem in a sense had disappeared with the droughty seasons following 1890 and while the millers took the grain, and the market for the hay and chaff was strong, the question was being asked– what was the practical value of what the scientist was doing? It was essential too that in departmentally stringent times to demonstrate that matters were progressing and should continue to progress.

During 1893 and 1894 the fledgling department had been in a state of flux. H.C.L. Anderson, the Director, had been retrenched in 1893 and replaced by W.S. Campbell, the Chief draftsman of the Department of Mines, and under Anderson the chief inspector of agriculture. The staff had been cut back and there were even threats from the Premier to dismantle the department altogether. December 4th 1894 was an important date for the justification of employment of scientists.

On that Saturday 70 or 80 influential farmers from around the district came, some from over two hundred miles away, bearing out the desire and need for these men to have information about new developments. The object of the Farm Field Day was for all the visitors to see the different varieties from all over the world and compare them. Nathan Cobb's intention in staging the field day was for the visitors to see the experimental plots and all the varieties, and to stress the practicality of the work being done. Small samples of the wheats in the plots were available and visitors could take some of those which they considered gave them the qualities they were looking for. The farmers would use their own judgement, being a compliment to the practical farmer where departmental men had been decried because they were not practical men.

The reporter noted that the local member of Parliament James Gormly, the Mayor and several aldermen and bankers were present. He did, however, note the absence of the Minister and the Secretary of the Department, but,

> *As regards the special objects in view –which were to afford an opportunity of inspecting the operations already carried out, and to listen to an explanation of future intentions of the farm authorities–they were not of course much missed.*

The occasion was chaired by James Gormly, and his remarks conveyed the air of pessimism about the future of agriculture for all those present,

> *Although agricultural prospects did not look very bright just now owing to the very low prices, still they should not feel discouraged. They had a suitable climate, fairly good soil for the production of grain, equal probably to any part of the world*

Dr Cobb followed with a detailed explanation of the operations of the wheat experiments and the objectives which were being pursued, improved gluten content, increased yield and grain which the millers would buy. He listed the varieties which their researches had found most suitable for the district: Televera, White Lammas, Hudson's Early Purple straw and Grosse's Prolific.

He acknowledged Farrer who was present, as the chief wheat experimenter in the colony and said that he could give them important information on the subject of the experimenting. Farrer however said he was not prepared to speak because he doubted whether anything he had to say on the matter would be acceptable to wheat farmers. He said that his experiments had been carried out with different objectives in view. He was seeking to develop wheats which possessed all the best qualities, *irrespective of other considerations, whereas the farmer had to grow his wheat for the market.* While he granted the value of the Purple Straw variety of wheat which was a favourite variety among farmers, he maintained his long-held view that the hard wheats were most free from disease.

The final speaker Mr Bennett, who by long farming practice in the district, was accepted as the father of agriculture round about, supported Cobb but reminded everyone that the farmers had to grow for the millers and Purple Straw was the class of wheat from which they had been able to obtain the best return for their efforts.

Although the matter of the wheat experiments had been in the newspapers view the most questionable activity of the Farm, the visitors could also inspect the newly planted orchard so that in general would have been able to go away satisfied and perhaps with an ear of wheat to try tucked in their pocket. The newspaper still took issue in an editorial in the same edition of the paper as the report of the day with reference to the experimental work,

> *It is of course of immense importance that wheat growers should be furnished with the variety of seed best calculated to arrest rust but it is still of greater importance in view of existing conditions that they should also be taught how to adopt a system of agriculture in which this cereal should play a less important part*

Despite the scepticism on the part of newspapers and reporters farmers were soon sending for samples of wheat which had appealed to them in the experiment plots in order that they too could trial them in their own paddocks. The letter books of the farm in those early days record many letters ordering particular samples from both New South Wales and Victoria and replies advising of their dispatch or alternatively saying that the supplies had run out. The sale of seed wheat was one of the ongoing commercial enterprises of first, the Wagga Experiment Farm and later of Cowra and Glen Innes.

HENRI DE VILMORIN

FOLLOWING THE initial Rust- In- Wheat Conference the exchange of letters between Henri de Vilmorin and William Farrer continued through the whole decade until the Frenchman's death in 1889, but continued with his son Philippe.

Farrer's correspondence with the experts of the time echoed his letter to Rev. W.B.Clarke in 1873 that it was difficult to get a book on his immediate topic and that Clarke was his book. This is marked in these letters from Henri de Vilmorin. This man and his father were regarded as pre-eminent in the 19th century for their plant breeding expertise by hybridising. The series of letters was written, no doubt when Farrer was spurred into gaining support for his methods from someone who was expert enough to add authority from outside Australia to his ideas

This letter was written from Verrieres le Buisson in Seine et Oise on December 1st, 1891 and is one from the beginning of the correspondence judging from the internal evidence. There had been one or two introductory letters during 1891

Dear Sir,

I have read with the greatest possible interest you letter and the extract of a paper on rust which you kindly sent me. I entirely agree with your views on the subject and consider it very fortunate that I am familiar enough with the English language to keep up a correspondence with you on the subject. Of course the study of the kinds of wheat most suitable to our western and rather moist climate is my principal object, but I am working in the same direction with you in so far that I have been asked repeatedly to name varieties adapted to the climatic conditions more similar to those prevailing in Australia than to any which we experience in France.

Some correspondence of my own are trying comparative experiments at the present time both in Spain and in several districts of Algeria. I am sorry that I cannot send you samples of the kinds of wheats which they are growing as I sent to them every seed I possessed in order to make the experiments as numerous as possible. Resistance to rust was not a primary object with me, as the countries for which those kinds are intended are not much visited by rust; I rather looked for very early and drought-resisting varieties. But those qualities and especially earliness is rather in favour of resistance to rust than otherwise, as it leaves less time for the invasion and spreading of the disease.

I saw that you ordered my 'Catalogue Agronomique' from the firm. This will make explanations easier and shorter.

I consider that you have most chances of finding wheat answering your desiderative in those introduced under the section 2,4,F,10, 26–B and 30.

The kinds which I would at first sight name as most likely to reward your trouble are the following:

Talavera de Bellevue

Richelle Blanche de Naples

Blé du Cap, a large Feuille

Blé de l'Inde Pietet

Blaue de Ogthere

Barbu Liataf

du Lazistau

Nongrois

Richelle de l'areute

Blé du Lizistau is the sort alluded to in Les Meilleuse Blé as the least liable to rust of nay wheat with which I am acquainted.

Thank you heartily for the sample of Leake's wheat. It came to hand just in time to be tested along with, although not in strict comparison being sown later, with my general collection.

Looking forward with pleasure to future communications from you, I remain,

Yours very truly,

Henri L Vilmorin

The following letter was written in March 1892

Dear Sir,

I am very much obliged indeed to you for the copy of the Report on the Conference on wheat which came to hand about three weeks ago and also for your very interesting letter and for the examples of wheat which we received lately.

I am very sorry I cannot write at length on the subject as I have been rather seriously ill for the last ten or twelve days and only began today to be able to stir at all.

The early kinds of wheat which you had the kindness to send me came to hand just in time to be included in my comparison trials of spring wheats. Of course you know that we sow our autumn or winter wheats in October or at times as late as January and sow spring wheats in March. Both kinds are harvested in July in Central and Northern France. I admire the size, beauty and plumpness of the seed you most kindly sent.

I expect to hear in two months of the observations made in Algeria (where the harvest takes place in May) and on the Riviera (where the wheats ripen in June) with some early varieties sent last autumn for comparison trials. I may send you some seed of such kinds as prove useful in those hot and dry places.

I have read with great interest of Dr Cobb's and your observations on the phys-

iological characteristics of a wheat that would be rust avoiding. I find however that on one particular my experience does not agree with yours. It is in the influence of the glaucous or bluish colour of the leaves or sheathes of the wheat. Blé de Moe which was with us nicknamed 'blé bleu' on account of the very marked bluishness of its foliage and stems especially at blooming time is one of the most liable to rust. So is 'blé de Bordeaux' which can scarcely be told from Blé de Moe when the heads turn brown at ripening time; it is also badly damaged by rust.. On the other hand the common 'Engrain' Triticum monococcum which is almost entirely free of rust under any conditions is a grass green wheat. You enquired about a 'Cape Wheat '......in the list I sent you. Now we have grown at different times two varieties of Cape Wheat. One is a bearded wheat, Blé du Cap Barbu. The other is beardless, a short stemmed stiff early white seeded wheat. It is the one we term "Broad leaf Cape wheat. It is certainly the latter I meant to suggest.

I did not trust myself to open the boxes of diseased wheats. I handed them closed to Mr Prillent who acts here about in the same capacity as Dr Cobb with you. Having been so... I did not hear about the results of his investigations as yet. By the Bye, how cleverly packed the wheat samples were, so that even if every paper bag was rotten which was the case, no mixture should occur. With every assurance of regard,

I remain, dear Sir,
yours very truly,
H.L. de Vilmorin

This letter may have been written in 1892, after Farrer had been to the Adelaide conference. It is dated May 17th.

Dear Sir,

I duly received your kind note of April 1st and shortly after the report of the rust conferences.

I am exceedingly busy at the present time and much to my regret cannot undertake to reply at any length. Yesterday or the day before I mailed to your address the translation published by the London 'Farmers Magazine' of an article which I contributed some months ago to 'Meuniere Francaise'. I much regret I could not lay my hands on a copy of the French periodical, as I understand that you read French well. It would have given you a better insight into my views as the translation by the 'Farmers Magazine' is a very imperfect one. I could not read it through but corrected in pencil some of the more conspicuous mistakes.

I sowed the samples of wheat received in the course of our winter, Did I say in my last letter that the heads of wheat put as diseased are not affected by any fungus but simply inclined to take a leafy development, to become what the botanists all 'virescent' the effect of the particular state of those heads leads to sterility, all the floral organs being transformed to chaff, but it is not catching. I sowed for experiment some seeds taken from the affected ears.

*The 'Catalogue synonymique des froments' is out of print at the present but
will be issued soon in a second and more complete edition.*

Believe me, dear sir,
Yours very truly
Henri L. Vilmorin

The following letter was written in September 1892, before Guthrie's mill
was in place and when Cobb was doing his chemical analysis of wheat.

Dear Sir

*I experienced some delay in obtaining an account of the mill intended for
preparing samples for analysis. But now I am in possession of all the information and I hasten to communicate it.*

*The apparatus was designed by M Aime Girard and works well. In order to
obtain not <u>flour</u>, but a fine powder in which the albumen and the skin of the
bran of the wheat seed are mixed, two successive operations must take place. The
seed is first ground through the revolving pieces of steel held at some distance
from one another and then passed again through the same so pressed(by means
of a screw) as to be in contact with one another.*

*The apparatus is made by Mr Digeon of Paris, engine-maker, Rue du
Terrage No.17a Paris. It is used only by laboratories, is not in frequent demand
and therefore remains rather expensive. The cost would be about £20 or £24 stg.
It should be ordered some time in advance as the maker does not keep it ready
made. It is not thought that more than half a dozen have been made, although
it answers very well and is also for making other vegetable substances than
wheat ready to be analysed.*

*As you anticipated my trial of Australian wheats sown in spring gave no
useful results. I am preparing to sow the same again in October. We had a very
strange season. May and June which usually are rather wet were quite dry and
exceptionally warm, up to the 22nd of June, when a moderate rainfall helped to
fill the ears, but the straw remained shorter than it ever was known to be in this
country.*

With every assurance of regard,

I remain, dear Sir,
yours very sincerely,
H.L. Vilmorin.

In the then stringent times of the Department of Agriculture and that it was
to was to be experimental this would have not been an acceptable acquisition. In
addition there were still eddying about many theories about the nature of wheat
which would a factor in reluctance to acquire the mill. Farrer was ever impatient
to have matters expedited so he must have wanted to know more about M.
Girard and where he could contact him,

Being on the eve of going to the Riviera for some weeks, I haste to reply to the queries transmitted from headquarters. M.Aimé Girard's Paris address is 44 Boulevarde Henri IV. M. Aubin's 366 Rue Saint Honoré.

M.Girard[16] published a pamphlet on 'The Structure and composition of wheat seed' which is much thought of here. I am not aware that M.Aubin published any thing in book form. But a good many papers and essays of his are scattered in the Bulletins de la Societé en Agriculteurs de France, and I presume in scientific periodicals.

I am very much obliged for the gift of the report so kindly sent to me and also for your considerable attention in having my name put on the distribution list for the Agricultural Gazette. I will certainly read it as punctually as I can, which is not saying a great deal as we are so terribly pressed for time in the old world.

I have also to thank you for the many samples of old and new wheats received in the autumn. I am touched by your confidence in sending me new crosses not yet out. I will study these with care and will report on them

I think I told you that the trials last summer were unsatisfactory. Spring and summer were so dry that some of the wheats were scarcely 18 to 20 inches high in the straw.

I sowed in October all the new varieties received and made a rather more extensive trial with the seed harvested from the former varieties.

For the last few days I have been preparing a paper on the wheat rust, in which Dr. Cobb's and your own investigations are duly mentioned. I shall make a point to send you a few copies as soon as it is in print – only it is in French, but I think that you can read it well.

You have guessed quite right that I often suffer from rheumatic complaints, only they are very variable in form. I find Mount Doré waters, with some time on the mountains a good cure and especially when I snatch a few days leave of absence, boating on the Mediterranean.

With best wishes and regards, I remain, dear sir,

Yours very truly,

Henri L.Vilmorin

Vilmorin, as were most plant breeders in the wheat growing countries in the temperate zone was interested in any of the work relating to rust, so that Farrer's package would have been of considerable interest.

On September 1st 1899, Farrer wrote to Vilmorin one of his long letters in his now rather crabbed handwriting. The letter shows a man confident with what he is doing.

My Dear M. Vilmorin,

I am forwarding to you by this post a sample of wheat which I think you would like to receive. It is one of my own making and is sufficiently fixed in type to have a name given to it. The name I have assigned to it I named Napier, after a town in new Zealand. The reason why I think you might be glad to receive it is on account of its high flour strength–viz. 68, which corresponds to about 329 lbs of bread from the 200 lb sack of flour, as flours which take up so much water in being made into dough only lose the same proportion of the weight of the dough in baking as doughs made from weaker flours lose.

I think Napier would be quite likely to prove suitable for your climate, owing to the fact that it has resulted from a cross between an almost pure fife wheat (a Fife with just a dash of the blood of Blount's Lambrigg) and your Autumn Saumur. Thus far I have only one cross-bred which has shown a higher strength than 68 (although I have two others which have shown that strength) and that is a Fife-Indian–one of the new class of wheats I'm making by means of crosses between Fife varieties and strong Indian sorts.

Napier, as you will doubtless see, is a variety with rather short and stout straw. I have selected for shortness of straw, but this year I have planted other strains of this cross and shall expect to find in the drills many plants with much taller straw, and that, in France would doubtless be desirable. The flour of Napier has one fault and that is that its colour is of too deep a yellow tint. This may not be considered a serious fault in France, where you appear to be less exacting in regard to the colour of the flour that is the case in this country.

It may be that Napier will not retain so high a flour strength when it is grown in France: but I think it is likely to be valuable for the proportion of flour strength which it retains. I have not sent Napier to anyone else in Europe, nor do I intend to do so, but I did send three years ago to Professor Sitensky 17 of Tabor, Bohemia, a small sample of the cross from which I have fixed Napier. Another crossbred(unfixed) which I am sending you, had as mother an unfixed cross-bred of identically the same parentage as Napier and Crepi for its father. This might give you a desirable variety and hardier than Napier.

In connection with what you have told me of the superiority of red wheats for the production of white flour, I am this season carrying on an experiment which I hope may give me an idea whether this arises from the fact that the colouring matters in the grain are concentrated in the bran of coloured wheats. I have planted drills with white seeds and with coloured seeds of the same cross bred parentage; this I have done with three of the different crossbreds. An examination(comparative) of the flours produced by the white and coloured grains of these crossbreds will give us an idea whether the colour has anything to do with the colour of the flour.

A serious fault which I find is attached to the flour made from some wheats is that it will not stand wetting– that is to say, if it be wetted, it dries with a grey colour and on that account makes bread with a greyish shade of colour. This is a fault which appears to be attached to several varieties in different degrees: early Baart, Quartz, Allora Spring, Steinwedel, a variety called in South Australia American Bearded, Inglis's Rustproof etc. among them. Of

European varieties Hunter's white is one which I have reason to suspect. Whether this is a constant fault, or is one which is dependent of the season, I do not yet know; but I have reason to suspect that faults of flour colour are worse after a very dry season. This is a subject of which it is difficult to get a certain knowledge, for the presence of dust in the mill–of even a very small quantity–is likely to mislead.

Since I last wrote to you I have been offered a position as wheat experimentalist to the Government. I continue to carry on my work here, but it is also extended now to the Government Experiment Farms, where it is carried on under my direction. This circumstance has caused me to alter my work somewhat and to devote less time to aims which are not of immediate practical importance. It is for this reason that I am giving up almost all the crossbreds which have been made with European parents on account of their being too late for our climate, and when I was sorting any wheats lately, I put aside a number of those I was rejecting which I thought would be likely to be worthy of attention by you. These I propose to forward to you in a few days and you will be able to give a trial to much of them as you care to.

After next harvest, also, I hope to be able to send you some varieties which may interest you. They have been collected by our Government on my recommendation from Russia(the Volga district) Asia Minor (Theodosia and Empatoria) and Persia; the varieties from the latter country ought to specially interesting. I have also a number of new varieties from India which are not the less valuable for having their native names.

I am anxious to recommend our government to import some new varieties of oats which are likely to suit us. In this country oats are grown for hay and for the grain, which is exclusively used for feeding horses and not for milling...

...After five seasons of drought, The present promises to be more favourable, but the critical time for the wheat crops has not come yet, and just now we are getting a dry spell.

With kind regards,

I am, my dear M. Vilmorin
Faithfully yours,
William Farrer

Vilmorin's letters trace the Farrer's individual line of investigation and account for some of the wheats which he imported, tried and crossed. In examining Macindoe's list of Australian wheats there are few of the many crosses from this source which remained in the Australian catalogue. By the time Henri de Vilmorin died in 1899 Farrer was about to establish his reputation and Australia's export potential with wheats which had their origins in many other places other than Europe.

FARRER'S BREEDING STRAINS

IN 1894 IN THE MIDDLE of a long drought, Farrer planted Improved Fife (itself a selection from Red Fife) was made by A.E. Blount and sent to Farrer. The aberrant plant(14A) with red straw was an earlier maturing plant than the rest in the row. Although Farrer does not appear to have identified it as Purple Straw, he as usual made a careful collection of the seed. The following year, 1895 he crossed it with Yandilla, a cross between Improved Fife and Etawah, an Indian wheat. Improved Fife had a pedigree going back to a sample of wheat which had come originally from Danzig, in North Germany. Vilmorin, in a letter to Farrer said that it was most probably grown there under the name of Juli Weizau (July Wheat). A sample had been sent by a friend in Glasgow to a Canadian farmer, David Fife of Ontario in 1842. Canadian and some American wheats had Fife as a progenitor, gaining for a class of wheat the generic name of Fife for world class grain. The Red Fife wheats became the foundation for some of Blount's varieties as it did in Canada imparting excellent baking qualities.

With the qualities of production from Purple Straw, the flour quality of the Fife strain and the drought and rust-resistance of Etawah, Farrer went on to produce a variety with short straw–not a hay wheat as the current wheats were– but a wheat suitable for the harvesting machinery then on the market. Farrer was not satisfied entirely with the grain he got from his Yandilla x 14a. from the standpoint of flour quality, but it appealed to him because of its plump grain. This wheat did not present the golden grain of the landscape painters but the waving grain was a bronze colour in the paddock.

By the time he was employed in the Department of Agriculture he had enough seed to conduct a field trial of Federation which took place on Iandra, a property near Cowra where share-farming was practised Farrer kept it to make a field trial of its potential in other ways, including the suitability for machine harvesting with its short straw at Wagga Experiment Farm. It was released in 1902 as Federation., honouring political Federation

When Federation proved so productive he discounted the praise heaped on it by pointing out its lack of rust-resistance and lack of strong enough flour for good baking quality. Notwithstanding his reluctance it was released to satisfy the increasing pressure to produce a big-yielding, drought-resistant strong-strawed wheat to suit the times and the needs of the farmers. With the changes in the Land Acts and improved machinery, land was being opened up for wheat under the government's closer settlement policies. Federation served a practical purpose.

Farrer's honesty would not allow him to claim that with Federation he had attained the holy grail of wheat breeding.

In his work of crossing to make Australian wheat for our climates and soils, Farrer drew on varieties in which he considered there were qualities which would work. Varieties from Professor Blount, Blount's Lambrigg, Blount's Fife and his wheats named after rocks and minerals, like Amethyst, Horneblende and Gypsum; from Vilmorin in France, Algerian Poulard, Aurore and Telavera; from India wheats like Etawah all added characteristics to the varieties which were to prove popular with farmers, millers and bakers. Nevertheless, until 1900, Farrer's crossing, growing, selection and testing in different locations tried everybody's patience. While his work may have drawn criticism, one of his main achievements was to show that it was possible to create varieties which suited particular needs of climate, such as Florence and Comeback.

It has been suggested that Farrer made multiple crosses haphazardly. The implication was that he was working on ideas based on ignorance. However, when he and Guthrie began to investigate the practical application the milling he could eliminate varieties which he had developed even at an early stage, when they did not display the important characters which he was hoping to fix. It added a new dimension to his method of working, making a careful selection of parents, using those which carried his desired characters. What he was doing was taking Blount's scanty description of making new varieties, a reading of Darwin and earlier plant breeders and a mathematical mind. The most important ingredient for Farrer and his progress was that he had time to continue over three or four years to breed a plant which he could declare fixed. Not all his fixed varieties met all his self-imposed criteria to fulfil his stated aim. Not even Federation was his fulfilled dream although at the time it met many of the needs of the farmers and millers more than satisfactorily.

In 1900 he had released Bobs, a variety of wheat which proved popular before Federation was released. Bobs was reputed to have been a cross between Blount's Lambrigg and Bald Skinless barley made by Farrer in 1896. Although it had many drawbacks such as slightly weak straw and poor tillering, it proved to be a breakthrough in the production hard grain of strong flour. Its naming was a compliment to Lord Roberts, the British general in South Africa.

During 1903–04 Farrer was much exercised that the seed which was being sold to farmers purporting to be Bobs but what he called 'spurious Bobs'. He tried to trace what it was and suspected that it was one of his discarded crossbreds, possibly emanating in New South Wales out of Bathurst Farm. The harvest from this seed was being rejected by the millers with, of course, financial losses to the farmers growing it.

Bobs was in a sense an unlucky wheat for Farrer as a batch of impure seed was sent from Wagga Experiment Farm to the Victorian Department of Agriculture. He, himself, had not parcelled the sample but it was done by employees on the Farm apparently with unclean equipment. There were complaints about the outcome of the Victorian sowing and Farrer's reputation was on the line and his disagreements with McKeown were further exacerbated. The whole episode lead him consider the conspiracies which were being mounted against him

In 1904 he sent a sample of Bobs to Mark Carleton, describing it as a rust-resister and that it had proved in the harvest before to be a good variety. He thought that Carleton would find it a successful spring wheat in the United States. By 1937 it was still proving to be a foundation breeding wheat on several experiment stations.

Federation overtook Bobs as the favourite variety in spite of the fact that from the point of view of the miller that Bobs produced a strong flour whereas Farrer had said that Federation produced weak flour. Federation's two other advantages were that: it yielded well and the short strong straw was ideally suited to the new harvesting machinery. Farrer said that Bobs was especially suitable to the cooler climates where Federation had a more universal adaptability, especially in the drier areas, excepting western Australia.

In spite of the caveat Farrer had put on the new wheat when it was released in 1901, in the April 1902 edition of *Agricultural Gazette* he wrote,

> *In the mill Federation produced a high percentage of flour of excellent colour and texture of appreciably higher strength than its parent Purple Straw. As Federation ripens at the same time as Steinwedel, and is much less rust liable and holds its grain satisfactorily it is possible that it may replace that variety with advantage as a producer of grain, bu the shortness of straw unfits it as a hay wheat.*

All the same, in spite of the pressure to release the wheat, it was not widely grown. It took until 1905 for there to be enough seed at the Cowra Experiment Farm to put it on the market but by 1910 it had become a widely cultivated variety and, among farmers and millers, it remained a favoured wheat until the end of the 1920s. For the farmers in the Wimmera it saved their industry, both Farrer and Federation being commemorated in Minyip.

Although it remained popular in most of the area of the eastern states, it was never a success in Western Australia. It was introduced into the United States where it was used on many experiment stations. It was commercially distributed in 1920 and on the Pacific coast it was an important variety. When Walter Scott Campbell, the ex-Director of Agriculture, returned to Australia he reinforced the value of it by recounting his visits to those states where it was extensively grown.

It was replaced in Western Australia by Nabawa. When George Sutton went to Western Australia, he took with him some of the unfixed varieties and Farrer's methods. From one of these he produced the pre-eminent wheat, Nabawa, named in 1915, which became one of the leading wheats grown in many other parts of the country until about 1938.

LONG DISTANCE ADVICE

WHILE COBB WAS working in the field and his laboratory, Farrer was at Tharwa writing lengthy letters both at home and abroad seeking advice and reports on the outcomes of his work and directions by which he could achieve his aims.

In 1921, a package of copies of letters which Farrer had sent to correspondents in the US and Canada was presented to New South Wales by Professor E.M. Shelton. had arrived in Australia from the United States in 1890 about to become an agricultural adviser to the Queensland Government. Before proceeding on his way he attended the Rust-in -Wheat Conference in Melbourne. It was about this time he met Farrer and became an admirer of the man and his work. Besides collecting and presenting the letters, he also gave an address on Farrer's work in the early 1920s. That he was an admirer of Farrer's can be judged in the following letter.

In 1894 Farrer had a letter from Mark Carleton, attached to the Division of Vegetable Pathology, saying that he had been advised to contact Farrer by Professor Shelton. Carleton wrote,

> *I am referred to you as a very proper correspondent in connection with my work on the wheat rust problem*
>
> *…I knew of your work in this line of course and intended to write to you before, asking if you would aid me in obtaining resistant varieties, but at the time we placed the whole matter of foreign varieties in the hands of Peter Henderson and Co of New York City…you would probably be the only one we would really need to correspond with in Australia.*
>
> *I hope that hereafter I may be considered one of your permanent correspondents.*

In 1894, always the letter writer, Farrer wrote to Professor Galloway of USDA and outlined what he had been doing at Lambrigg and beyond:

> *I may state that it is now twelve years since I began to grapple with the rust problem. At that time I was unable to take the matter in hand practically and my efforts were confined to making public my views and offering suggestions through the press. It is eight years since I began to make practical experiments for myself.*
>
> *The first suggestion I ever ventured to make on the subject was one which has since yielded excellent results. It was that farmers who had rusty crops should make search for individual plants which were free from the parasite and*

thread the plants which are the first in each lot to come into ear, taking field notes and in making a casual cross.

Later in 1898 he was to read a paper to the Australasian Association for the Advancement of Science in which elaborated on his whole procedure. Guthrie commented in a memoir on Farrer in *The Lone Hand* after his death that it was not well understood at the time and that his listeners regarded him as *an eccentric visionary.*

Farrer continued to correspond with Mark Carleton working as a wheat-breeder from the mid-west of the U.S. and the farmers there were having trouble with rust infesting the wheats they had brought with them from Russia. Carleton continued to write to Farrer about the work he was doing. From the beginning the letters amicably discussed the methods they were using and results they were getting. They exchanged publications and comments, Farrer respecting the knowledge Carleton displayed and the findings he was making. In 1902, Farrer described an experiment he had conducted on bunt infected wheat and the effect of the infection on the milling quality of the resulting grain. As always from 1892 he was concerned not only with the yield but with the production of flour of suitable quality. His conclusion from the experiment showed that there were no differences in flour quality between the bunt -free and the bunt-infected varieties.

Farrer, discussing his success with the *efforts I have been quietly and systematically making to breed wheat of high flour strength,* moved onto the wheats which Carleton had obtained after the American had *scoured the world for wheats* and asking him for samples of some of the varieties, more especially suited for the dry parts of the country.

Cobb's work on nomenclature too came under criticism.

I dare say you have seen from our Agricultural Gazette that Dr Cobb attaches great importance to our having a universal or uniform nomenclature fro wheats. I do not agree with him in regard to the importance of this matter for the reason I think that, in the near future, improved varieties of wheat will follow one another almost as quickly as at present do new varieties of many of our table vegetables, and that the duration of the usefulness of a variety, until it is replaced by a better, will, in general be relatively short, and because the external characters of a variety of wheat have not the stability or the importance that have the characters of a natural species or a natural variety.

I think that, instead of spending money and effort in securing uniformity of nomenclature, an Agricultural Department can do far greater good by making, for its farmers, new varieties possessing the qualities which give real value to a wheat. An opinion that I hold is, that it is the duty of the Agricultural Department is not merely to look after the interests of the farmers, but, inasmuch as it is maintained at the public expense, those of the public as well, and one respect in which the interests of the public can be looked after, one in which I am trying to look after, is by making it a sine qua non of the wheats I make for our department, that they shall, as regards their gluten content and flour-strength, reach a high standard of nutritive value. I think that an Agricultural Department ought to do what it can in the way of putting into the hands of its farmers, and encouraging them to grow exclusively, wheats which will furnish its people with as nutritious food as possible, as will as be safe as possible to grow by being resistant of diseases and of the injuries which the climate is apt to inflict on them. Of course in order to do this, an Agricultural Department has not only to carry on the making of new varieties in a systematic manner, but to be possessed of a small testing mill, by which their wheat makers may know exactly what they are doing.

These views may not have been entirl;y restricted to Carleton letters. In 1900, Farrer had written wryly to Kahlbaum that,

The fact is that I am in disgrace with the department, because I am such a disturbing factor. I want to see progress made and they resent this. I fear my influence there just at present is very small indeed. But I do not mind that.

Farrer continued his letters to Carleton each following the others progress on the improvement of wheat. In every country the problem of growing population through industrialisation and the likelihood of war in Europe concentrated government efforts to improve the food quality of the farm produce.

In Canada, improvement was moving ahead on the western plains where a series of experimental farms had been established to deal with agricultural problems, the most pressing of which was to find a wheat which was early-maturing and so able to be ready to harvest before the frosts made their appearance. Dr William Saunders became the first director of experiment farms.... Introducing varieties from the US and Russia. William Saunders had two sons who involved themselves in wheat breeding. One, Percy, developed Marquis, a variety which became the benchmark for all Canadian wheat. The other was Charles Saunders, who like Farrer and Guthrie and Cobb was interested in determining the gluten content of the grain. He had the 'Chewing test' which he applied.

By 1905 he was exchanging letters with Farrer on their common interest in flour quality and gluten content. The question which he posed to Farrer was about the methods which the Australian researchers were using to get a larger loaf from a flour. Farrer wrote,

...we are not able to answer and I did not like to do so until I had enquired at the laboratory at Sydney. The fact is, that we only have a single small testing mill and that of an old out of date pattern...and will only deal with small amounts of grain. The quantity we are accustomed to grind for the testing is

save the seed from such plants. I suggested that by doing this they might become possessed of strains of existing varieties which possessed superior power of resisting the parasite.

This apparently harmless and reasonable suggestion was commented upon and ridiculed by our leading agricultural newspaper and it was the further thought I gave to the subject when I came to fight my battle with the newspaper in defence of my suggestion that lead to the formation of the opinion I have since held and acted upon.

During 1894 he advanced further along the path he was treading which was widening his horizon as to the possibilities which lay within his cross-breeding work. He began to work towards a product which contained within it the qualities of strong flour potential, good milling propensity, and heightened disease resistance, all combined with high productivity values. All these were to be combined with an ability for the plant to flourish according to the climatic environment for which it was intended.

Later in 1894 he again wrote to Galloway. This time jubilant about the assistance he was getting from the department. *Thanks to the co-operation of our Department of Agriculture which allows any analysis, I may think necessary for my work, to be made by the departmental chemist.* This was the beginning of the combined efforts of William Farrer and F.B.Guthrie. He continued,

Mr Guthrie …is now devoting himself to an examination into the qualities which cause one flour to be superior to another for bread-making. At present he is seeking for guidance data for a systematic investigation from the examination of some seventy wheats I have put into his hands…I may state that Mr Guthrie is furnished with a small roller mill by means of which he can and does make flour which is equal in quality and appearance to that which can be made from the same wheats on a large scale.

Guthrie was to report that with the aid of the pair of roller mills that a system was devised that allowed him to produce a flour *in all respects identical with that of the best millers.* He commented on the value of the experiment; it *gave him information as to the milling qualities of the grain which he could not otherwise have obtained.*

As well as writing to Farrer at Shelton's suggestion, Carleton was also recommended to Farrer by Professor Galloway as doing some interesting work. In a letter Farrer explained his routine to the young American:

…I did some cross-breeding. My routine was to work for a couple of hours before breakfast, marking by means of a black tie the plants of each drill which were earliest to show ears; between breakfast and dinner was devoted to making crosses, in the afternoon right up to sunset I spend time again in tying with black

only 12 oz. So that we only get on an average under 9 oz of flour and nothing of value can be done with that amount for elucidating the points on which you enquire. Mr Biffen of Cambridge has asked me much as you have done re the same subject is engaging much attention there. I would suggest to you an exchange of conferences with him...

We have thus far to all practical purposes confined ourselves to estimating the likely strength by the amount of water absorbed alone...So far as our experience goes and it has been pretty well confined to the three varieties Bobs, Jonathan and Comeback, the results of our estimation of flour strength have been borne out often as not by the experiences of the bakers who have dealt with such flours...The wheats grown here in 1904 which have been duly milled have proven to be excessively rich in gluten as well as of high strength (absorption of much water).

Farrer had been using different applications of fertilisers to improve the flour and gluten quality, but as usual he was not prepared to make any definite assumptions before there had been systematic experiments. On this head he suggested to Saunders that this would be a good avenue of experiment to pursue. Farrer by this time was feeling the strain of his work and was more inclined to give others leads for some profitable work. In 1907, following Farrer's suggestions and correspondence with Biffen, Saunders used the methods of grinding the grain and conducting baking tests to find the wheat which gave the better loaf.

Charles Saunders was knighted for his work.

THE SECOND FIELD DAY

Late in 1897 the Second Field Day was used to announce the opening the new laboratory the first specifically built Government laboratory and designed to Cobb's specifications by Walter Liberty Vernon. Vernon was the Government architect at the time and for that day incorporated some innovative ideas onto public buildings such as Court Houses intended for the hotter parts of the State, such as Bourke and Wagga Wagga, both designed in 1899. The new laboratory was designed in 1897–98.

Before the occasion a *The Wagga Wagga Advertiser* journalist, in October 1897, reported on a visit to the Farm to see what progress had been made. He reported in detail the interview and inspection tour he had with Dr Cobb, at that time the Acting-manager.

The article had all the appearance of recording the ideas and comments which Cobb was always able to communicate to the press. Commenting: *As a preliminary it may be stated that in the establishment of the Farm at its inception a somewhat mistaken policy actuated the authorities in the construction of the buildings. The cause of this policy being inaugurated was undoubtedly the difficulty the Minister for Agriculture experienced in obtaining from Parliament a sufficient vote to thoroughly establish the farm with complete and permanent buildings.* To Cobb the makeshift, while he always worked his way around it, was not something to go without remark. It was in his report of 1897 that he said he was pleased to see that now officers of the department no longer had to work in tents or in a bough lean-to.

Cobb's first laboratory had been in a tent, then succeeded by the weatherboard building Cobb persuaded the Minister to allow and which was put up in a smart fashion before the first field day in 1894. The journalist in the course of his article contrasted this building with the new laboratory, *situated right in the middle of the experimental plots, instead of, as in the case of the present one, at a distance of over half a mile.*

By the time of the journalist's visit, the new brick building had been built on the eastern side of the hill. It was the first proper laboratory built by the Department and designed for specific investigative and teaching purposes. *The building is of two storeys and is in the form of a cube, this being the shape which least exposes it either to the cold of winter or the heat of summer. It is substantially constructed of brick and round each storey is a wide verandah or balcony.* All four rooms were accessed through doors which opened onto the verandahs. The foundation of the building, granite blocks and concrete, was also divided into four. To further facilitate the cooling and ventilation of the building, a shaft rose through the building in the manner of a chimney and opened into each room. It is impossible to tell from the plan whether the vents into each room were also fireplaces, but in the winter cold it is logical to conclude that this would have been the case.

It was planned to be labour saving for the scientific staff and as *perfect as human ingenuity could make*. The description of the building was fuller in the newspaper than in Cobb's own description in his 1897 report: *Each of the two floors contains 4 large and well-ventilated rooms, the access of which is from the outside of the building but, believing that it may be necessary to provide for direct communication from room to room, the walls have been constructed in such a way as to allow of this being affected at small expense. Each room has been specially laid out for its intended purpose. Thus the room occupied by Dr Cobb has in it steel railway rails set in solid cement in order that the microscopes and other apparatus may be placed upon them in such as way as to provide against vibration. The photographic developing room has also been built with all possible convenience for the carrying out of the art and other rooms are equally adapted to their utilitarian objects.*

The top of the building is flat with a high brick rail and cemented floor. As an awning will stretch over the roof, upon its solid and crackless floor may be conducted the many operations of sorting and sifting grain…Each of the upper landings has a sliding panel in the rail so that goods may be easily hoisted from the ground. It excited the journalist to be ecstatic over the extensive views right to Mount Kosciusko from the roof.

The equipment of the rooms repeated a pattern Cobb had developed over time. An illustration in *The Sydney Mail* at the time of the Bong Bong investigations and others later in the *Agricultural Gazette* give an idea of how his arrangements were made. In Frieda Cobb Blanchard's biography of her father she describes his laboratory in Hawaii and it could well be a description of the building at the Wagga Experiment Farm

The granary, or elevator, as Cobb call it, or 'Cobb's Folly' as others called it, was close by down the slope of the hill. *The situation is below and to the right of the new laboratory and the building is upon the most modern lines.. The granary is of three floors and for purposes of inspection, as for use, one can best commence at the top. The site of the building has been cut away from the hillside, and on the hill at a level with the top storey will run a roadway. Across the intervening space will be placed a bridge and over this bridge will be driven the carts containing the grain. The grain then be released from the carts will fall upon the floor. Along the floor at regular intervals are trapdoors leading to bins of the shape of an inverted triangle. The bins open with a small slide into the chamber beneath. On this floor (which also communicates with the roadway in case the upper floor is full) there is large high table upon rollers, which can be rolled beneath each bin as is required. The operator will have upon the table a sifting machine to separate and cleanse the grain. Beneath this on the floor will be a similar machine and the grain when cleaned can be bagged or allowed to drop through other bins (at which, if necessary the cleansing process can be repeated) to the bottom. At the sides of the floor are openings through which the dirt and rubbish may be swept…The building is of the most solid character, and here, as at the laboratory, all possible care has been taken to provide for the saving of labour and to add to the convenience of the work. The principle upon which the building is erected is the elevator process…Except for the primary work of carting the grain all the cleaning will be conducted on the principle of making the grain by its own weight become its own carrier and distributor.*

The journalist was not of the Ernest Hemingway school, so short paragraphs were not for him.

The account of this exploratory visit extended over two editions of the paper. The second part of the account of the visit was of an interview with Dr. Cobb, largely concerned with his ideas about how agricultural education should be carried out. These were basically those which via the *Daily Telegraph* were reproduced in the *Journal of the Western Australian Agriculture* in 1906. That Cobb, a public servant, should give an interview relating to what could be seen as a comment on Government policy would not to-day be allowed to pass.

What Cobb had to say in many ways came out of his own experience in his growing up and being educated. He was of the mind that access to the Farm Schools should only be available to those who intended to be involved with agriculture. To maintain an economic balance sheet balanced, he also felt that parents should pay and not see enrolling at a Farm School was a cheaper way of maintaining their children than keeping them at home or sending them off to a school charging higher fees. He maintained that the wrong class of student was being attracted who had for the most part no intention of pursuing agriculture. Class in this context was not in Cobb's usage social but category of work and he believed if the government gave concessions to one category it should be given therefore for the training of *shoemakers, tailors or other skilled tradesmen.* It was his opinion that the problem with agricultural education was that it was failing to produce agriculturalists: they became instead, *teachers, lawyers and merchants.* They were being educated to feel that they were above personally tilling the soil. Students were told about agriculture instead practicing it. They were returned to the country with the ability to carry on and improve agriculture without the desire to do it. They had not been taught the habits of work; not allowed to learn the many things necessary by constantly practicing the skills.

To allow students to enrol who might find the fees beyond them he proposed that they could earn their keep by working on the farm or tutoring those students who were not quite academically able. A further suggestion was that when they had completed part of their course they should drop out for a year to earn some money to complete the course.

While he agreed that the Farm Schools should be self-supporting he also believed that experimental work should be funded from general revenue because of the good which would accrue to the public from the outcomes of this work.

These opinions were the first to be voiced on the matters of agricultural education. Guthrie's views were published the next year in the *Agricultural Gazette,* in contrast to Cobb's views, which were for the newspaper, both at this time when he was acting manager of the farm and later in *The Daily Telegraph.* Cobb knew how to take his ideas to a much more influential medium.

Following the visit of the journalist, the stage was set for the visitors in November. The reports were part of what could be called an advertising campaign. The field day was widely advertised in the Riverina papers and was to extend over two days; farmers and others interested were invited to come and see what was being done and have the methods of experiments explained to them. It was an important occasion to which the Under-Secretary for Agriculture in the department of Mines and Agriculture, the Members of Parliament for both Albury and Wagga Wagga and the Mayor of Wagga Wagga came. The Vignerons and Fruit-growers Association of Albury as well as representatives of farmers organisations from Henty and other agricultural centres also came. The experiment farm now could widen the range of what was to be seen other than wheat: fruit, and vines as well as wheat. For the Government's wish to widen the kinds of rural products which could be cultivated to provide an added economic strength along with wheat.

The Under-Secretary was met at Bomen Station off the overnight Albury Mail and driven to the Farm but, for about twenty people who were staying at Bellair's Hotel a five-horse drag was provided. When they reached the Farm gate on the Pine Gully Road, a procession of vehicles formed and all made their way towards the eastern boundary of the Farm where the orchard and the wheat experiment plots were. Cobb stood on the box-seat of the drag and explained to the visitors the extent of the Farm's operations.

The new laboratory was built on the side of one of the hills on the farm called the two Sisters. The next stop for the visitors was on top of the 'Sisters' where *unfortunately the clouds of dust in every direction detracted greatly from the clearness of the prospective scenery.* Nevertheless, from here they could see the thirty acres cleared for wine grapes and olive trees, twenty for grapes and ten for olives. Below them were the, Wheat experiment plots. Cobb had made every effort, beginning with the *Wagga Wagga's Advertiser* interview a week or so before, to promote the value of the form in the light of growing criticism about its value. It was, as his daughter later wrote his *touch of Barnumism.* He knew from his early days in Sydney, selling Cashmere Soap and Waterbury watches, the value of the view over the written word.

The morning must have been very hot as well as dusty, one of those precursors to a long hot Australian summer. The journalist described the weather as oppressive and was quite pleased when it was time to break for lunch, held on the ground floor of the new granary. About 130 sat down to lunch, hosted by Nathan Cobb. There were the inevitable toasts and responses.

In the afternoon everyone drove down to the western end of the farm to inspect the 200 acres of Allora Spring, a controversial variety of wheat, which was inclined to be one of Cobb's favourites but which Mr White of Belltrees near Scone had written complaining about its milling qualities. Cobb in its defence, wrote to say that Sydney millers had found it worked quite well when combined with other wheats.

The whole programme was repeated the next day.

THE MAKING AND IMPROVEMENT OF WHEAT

IT HAS BEEN SUGGESTED that Farrer made multiple crosses haphazardly. The implication was that he was working on ideas based on ignorance. However, when he and Guthrie began to investigate the practical application the milling he could eliminate varieties which he had developed even at an early stage, when they did not display the important characters which he was hoping to fix. It added a new dimension to his method of working, making a careful selection of parents, using those which carried his desired characters. What he was doing was taking Blount's scanty description of making new varieties, a reading of Darwin and earlier plant breeders and a mathematical mind. The most important ingredient for Farrer and his progress was that he had time to continue over three or four years to breed a plant which he could declare fixed. Not all his fixed varieties met all his self-imposed criteria to fulfil his stated aim. Not even Federation was his fulfilled dream although at the time it met many of the needs of the farmers and millers more than satisfactorily.

William Farrer had first enunciated in his letter to *The Queenslander* in 1882 a probable line of action. From this time he was in contact with Joseph Bancroft and writing to Professor Blount in Colorado and with Henri Vilmorin in France.

By 1890 he had developed his line of experimentation, which he enunciated in a letter to the first Rust-in-Wheat Conference that year. He did so again in a letter to the *Sydney Mail* in reply to a letter in that same paper in which the writer, Mr Smith Ellis, wrote that he had a cure for rust which he was willing to sell for £10000. In 1892 he went to the Conference of that year in Adelaide and confidently told of his experiments, and by 1897, just before he joined the New South Wales Department of Agriculture, he could clearly articulate the detail of his methods in a paper presented to the Australasian Association for the Advancement of Science. The next year the paper was published by the Department as *Miscellaneous Publication No.206*.

In his opening remarks he spoke deprecatingly of the history and progress of his work,

> *…I find my notes have been kept so unsystematically and that so much has been forgotten which is of principal interest at some stage of the work, that it would be extremely difficult to write a paper on that subject which would be satisfactory to myself or interesting to others.*

He qualified what he would discuss–the work itself:

> *After I had grasped the fact that it was its variability in regard to the amount
> of resistance offered to rust which rendered it possible to increase that quality in
> the wheat plant, I saw it would be necessary to observe in what other respects it
> varied before I could take in hand the making of other improvements.*

Farrer in his paper of 1898 said that the principle on which he based his
work was simple:

> *It is merely that whatever well-directed attention is given to the improvement of
> a domesticated plant in any quality in which it is variable, that quality can be
> increased and developed to an indefinite extent.*

Farrer, while he was interested in the innate qualities of wheat, the suitability of the wheat for successful and profitable milling was an important factor in
the end result of his experiments for farmer, miller and consumer. The emphasis on the milling quality would have come from his acquaintance with the business of buying and selling grain, conducted his great uncle William and his uncle
William.

From the consideration of rust resistance which he had originally made his
concern, he turned to other qualities of the wheat plant which could be
improved: the amount of gluten and strength of the flour, so necessary for successful baking; the suitability of varieties of wheat for the varying climates and
soils of Australia; the improved yield of the crop. He had qualifications about the
use of selection alone; the improvement could be short-lived and unstable in
nature. In his search for the desired qualities he widened his method from solely using selection of individual plants to the added use of cross-breeding to arrive
at a suitable outcome. This was a simple concept but his description of his practices revealed the meticulous method of work, observation and recording which
he laid out. Conducted on an area from 2–3 acres, and area divided into half to
be used in alternate years, these practices had carried him determinedly forward
by slow degrees,. His method had added value as the soil was poor and shallow
allowing him to grow his experimental wheats under conditions similar to the
drier parts of the country, where he said lay the possibility of further expansion
of wheat growing.

At a time when wheat was still mostly broadcast sowing, his planting by
hand, his seed set out in rows, each seed being carefully put in about 5 or 6 inches apart (about 10cms.) There was a distance of 16 inches between each row.
The length of each row was about 15 yards, (about 4 metres) containing about
100 seeds. Each day he walked up and down his rows, looking closely at each
plant during its whole growing period, making notes relative to the qualities he
was looking for, working to produce his ideal variety. This was his practice almost
every day of his experimenting life up till his death in 1906. At every stage notes
were taken with regard to the colour of the foliage, width and character of the

leaves, habit of growth, whether erect or prostrate or creeping, n regard to stooling qualities, and, in the case of unfixed cross-breeds, in regard to individual plants, and any that are remarkable should now be taken.

He considered that these notes would come to be of importance later on as the trials of the various crossings proceeded. In particular he was marking plants for earliness in heading, as he had concluded that, in the hot dry climate such varieties would be of more value than the late ripening varieties. At this point in the growth of the plants he marked the plants for the different qualities they possessed by tying narrow strips of light cloth to the stalks making sure that the same colour was used for the particular quality. He used black for earliness in heading. In each drill there were usually two or three plants which were marked–some with two ties, or even in special circumstances, with three. Everyone in the district could see this part of his work and soon labelled him as a crank–even a faddist.

It was this meticulous physical application of his time which became the most tiring for him. The crossings were better done in the early morning for finding the best pollen and then he would continue working until about mid afternoon. As the heat became greater so did the speed at which he had to work. He maintained that,

> *The work of making crosses is not difficult, but practice is needed in doing it quickly; and the eye has to be educated to recognise when ears are in the right stage to be impregnated and especially to see when the anthers are ripe enough to furnish good pollen. The getting of good pollen takes more time than any other part of the work.*

Continually throughout the paper he returned to the nature of his work notwithstanding regarding things like the inspection for rust as being tedious. He described his harvesting procedures, which at Wagga Experiment Farm, McKeown dismissed as being of no practical value to farmers and about which the Manager was always reluctant to take men off general farm work to do this job for Farrer. It would have been irritating to McKeown because, to Farrer *the harvesting …is work which cannot be hurried.* This is Farrer's account of his procedure,

> *The marked plants have to be taken first from the drills; in this the rule of harvesting each marked plant by itself, and of tying it into a bundle of its own must be followed strictly. Care must be taken in harvesting these plants to push all the marks up the stalk to the ear before the plant is cut. Plants from the same drill even if they have been marked in exactly the same manner for the possession of the same qualities in exactly the same degree must always be kept separate, for they may differ widely in their power of transmitting some or even all of their qualities. This rule applies even to the marked plants of a fixed variety; and with, of course, much greater force when they are unfixed crossbreds...*

> *The rule, therefore, of harvesting all selected plants by themselves, especially*
> *when varieties are being made, is of the very first importance, and on no account*
> *should ever be departed form. It is because I consider its importance so radical,*
> *that I have dwelt on it at such length.*

Following his explanations he reiterated the importance of keeping records of all that was done. The paper was a very tightly packed body of information, like the fruit in a Christmas pudding, and possibly as indigestible on first hearing. On the other hand his audience belonged to a generation when lectures and long sermons were the vogue.

In discussing the improvements possible in wheat he referred back to increase in the sugar content of sugar beet made by Henri de Vilmorin's father, Louis, when Napoleon offered premiums for the manufacture of sugar from this plant. It had originally run at 5% but, by what he denoted as 'intelligence, ingenuity and enterprise' the level had lifted to 21% producing what at that time could be considered as a luxury, but was only a starch with small nutritive value. To him, in the same way, intelligence and enterprise could produce the improved wheat providing the gluten which is superior to starch as a food, which together with increased starch in the flour could produce a nutritive loaf of bread.

Having laid the foundations of his approach to the whole business of developing new varieties, he discussed the product of the wheat seed itself, the flour and its potential ability to provide a good light loaf of bread drawing attention to the influence of gluten, the *lean-flesh-making constituent,* on the "strength" of flour thereby to achieve this aim. In doing this he credited F.B. Guthrie with the discovery that the differences in the bread-making qualities of the flour from different varieties of wheat were due to the proportions of the glutenin and gliadin making up the gluten in the flour. Guthrie had taken account of the investigations of Osborne and Voorhees at the Connecticut Experiment Station which had determined that the gluten was in fact made up of two constituent, glutenin and gliadin. From these findings of Guthrie's, other researchers in Europe and America tried to find out what proportion of glutenin to gliadin making up the gluten allowed the flour to take up most water to make the lightest bread.

This was the point where the co-operation of Farrer and Guthrie combined their efforts and found that the wheat plant was variable in these qualities and so could be modified and so possible to develop varieties which could produce the most nutritious flour. Farrer's philosophical approach to wheat breeding was not just for the high yield and economic gain but to provide bread of benefit to the populations of the world. That he should pursue what would benefit to his fellow citizens ahead of monetary gain would have marked him off as almost dangerously radical. In his paper he outlined what he called his dream. He said that in the past where the population had outgrown its supply of bread (he had lived through a century of revolution), 'fierce struggles ending in death' cut the population by what could he called natural selection. In his view, when the then current increasing population outgrew the world's resources for growing wheat, then efforts would be made to increase the bread making capacity of wheat so keeping pace with the increase in world's population. He went even further with the prediction that when the capacity to improve of wheat grain had been reached then the chemists would come forward with synthetic means of making bread form the simplest materials, so maintaining the nutrition necessary for the people.

In his quest for good flour he returned to his long-held views about the improvement of gluten content could be improved with soil improvement by the addition of nitrogen, not through chemical fertilisers but through leguminous plants especially clovers. He had, in the past, in one of his letters to the newspapers, drawn attention the value of these plants whose,

> *…nitrifying organisms will do for us in the soil itself after we have got to understand them, and how to create the environment.*

It was not until the 1930s as the quality of Australian wheat declined that his theories about that link were put into practice with subterranean clover..

His paper sheds light on his approach to his work, emphasising the fact that he had opened up a completely new concept of developing new varieties which remained unchanged in wheat breeding until Albert Pugsley and Ron Martin in the late 1940s.

In the preface of his 1873 pamphlet, *Grass and Sheepfarming* Farrer wrote;

> *This paper does not pretend to give much information that is immediately valuable for practical purposes; if, however it proves suggestive, and sets intelligent, practical men thinking, it will have served its purpose: it will do better still if it provokes useful discussion.*
>
> *The statement of individual observations and conclusions which have by no means been as yet accepted as facts, necessitates the somewhat frequent use of the first person pronoun: a fearless use has, accordingly been made of it.*

Farrer had by 1898 been able to take a less deprecating line, sure now that what he had to say had great authority. His paper needed no excuse for its presentation or content.

He once commented wryly, *difficulties, I know, are often ignored in dreams, although on the other hand they are apt to wax over large and become nightmares,* reflecting, no doubt, his own experiences as worked for over twelve years to produce a wheat, "Federation" which would make him famous.

Guthrie commented in his Memoir of Farrer in *The Lone Hand* in 1906 that this paper was not well understood when it was delivered and that his listeners regarded him as an eccentric visionary.

FARRER EMPLOYED

WHEN COBB APPLIED for leave to go overseas in 1898, W.S. Campbell was Chief Inspector of Agriculture and D.C. McLachlan, the Under-Secretary for the Department of Mines and Agriculture, approached the Minister, Sydney Smith, to have William Farrer appointed to the post of Wheat Adviser. Smith, briefed no doubt by Campbell, submitted the proposal to the Public Service Board:

> *It is now recognised that this Colony offers perhaps better natural advantages for the growth of wheat on an extensive scale than any other part of Australia, and this department has been doing everything possible to foster and encourage the extension of the area placed under crop.*
>
> *…In furtherance of this the Department has for some years past conducted very extensive experiments in the selection, crossing and growing of wheats…Hitherto the experiments have been carried on in a highly scientific and satisfactory manner by Dr. Cobb, but his absence from the Colony makes it necessary that someone else be at once appointed to continue and enlarge upon the experiments that have been made.*
>
> *This is still more necessary in view of the large areas suitable for wheat culture in the western portion of our Colony where until recently it was held that it was impossible to grow wheat at all. Recent developments, however, have shown that our western lands are as well adapted as any other for growing this crop on a large scale.*

Sydney Smith, in his submission, went on to make a more precise outline of the problem.

> *…It is important that we should experiment to ascertain what description of wheat is the best rust-resistant and at the same time posses good milling qualities, as well as ability to withstand the heat and sudden changes of temperature of many of our inland districts, and also to prove what variety obtained either by selection or by crossing, will best resist the droughts or partial droughts of our western interior. Investigations are also required to be made as the whether deep or shallow sowing of the various varieties will best suit the peculiarities of the climate and soils of our vast alluvial plains in the Riverina and stretching cross the south-western and northern-western districts.*
>
> *…I would strongly urge that Mr. W. Farrer who has made this question a life-long study and has achieved considerable success in this direction, should be approached and asked if he is prepared to continue the work commenced by the Department, with a salary of £350 per annum.*

Archer Russell in his book on Farrer considered the salary offer parsimonious, but it was on the higher end of departmental scale at the time, being only a little lower than that of a Farm Manager and equal to that of the botanist and the entomologist. Farrer himself said that the amount of money did not interest him. For Farrer in 1898, this was now an opportunity to be part of the government process with access to the experiment farms where he could trial out his new varieties over a range of soil and climate funded by government money and enabled him to have added secure money coming into Lambrigg as Leopold de Salis' financial affairs were coming to a head. In 1898 he was declared bankrupt, losing his seat in the Legislative Assembly. From 1894 George de Salis, his wife Mary and their children, as well as Nina's brother Harry had been living at Lambrigg. Now he would need more than the annuity from the Farrer Trust and his rent from his estate in England to sustain the property. Campbell, in 1912 in an address to the Royal Australian Historical Society outlined the Department's rationale:

> *In 1898 an opportunity occurred to appoint Farrer Wheat Experimentalist to the Agricultural Department, where it was thought his scope would be extended and his opportunities improved for carrying on the work which he had hitherto performed privately and entirely at his own expense in a most generous and unselfish manner, giving away to anyone he considered worthy the results of a vast amount of thought and also of expense. The appointment, it was considered, might be some little acknowledgment and recognition of his efforts.*

The appointment was made in August 1898 and Farrer wrote the following letter of acceptance:

> *…You offered me the position of Wheat Experimentalist at a salary of £350 a year, I have the honour to say that I am prepared to accept it on the terms which were verbally agreed to by Mr Smith in an interview I had with him on the subject earlier this month:*
>
> i   *That the position should be a permanent one and in no way affected by Dr Cobb's return.*
>
> ii   *That I should be at liberty to carry on my wheat improvement work by methods of my own choosing and that it should be done on Government Farms under my own supervision.*
>
> iii   *That I should carry on the wheat-breeding work at Lambrigg.*

By making one of the conditions *'methods of my own choosing'*, problems arose which were to do with his wheat experiments on the Farms where the Farm managers saw themselves as having no control over his operations or requests.

As he himself went on to write in his letter,

the extension of this work(wheat breeding at Lambrigg) to the different Government farms and the fixing at them of new varieties from seed of selected cross-breds which would be furnished from Lambrigg

…In addition to improvements in the wheat plant itself, it is, I consider, of even greater importance that I should conduct experiments for the purpose of ascertaining the methods of soil-management which are most suitable for our climate and the conditions under which our wheat growers are working. It has long been my conviction that the system of alternate cropping and fallowing which is being followed so generally by the wheat farmers of Australia is a mistaken one…it will be my endeavour to work out a more rational system of soil-management

He went on to outline how he would examine the way in which sub-soiling could be made to take the place of irrigation: leguminous plants could be trialed for green manure; soil nitrification explored.

Any one of these explorations and his attitude to routine would bring him into conflict with a manager like Maurice McKeown and his allocation of workers, machinery and space. Added to the likelihood of conflict with regard to the carrying out of his proposals was the fact this was not a junior employee coming into the department but a man who was older than any of them and who had from his schooldays been in a position where he gave the orders and had worked independently

The Minister had accepted Farrer's conditions but a letter sent to the Farm managers by the Under-Secretary appears to give Farrer independent authority overriding the managers:

I am directed to inform you that Mr William Farrer has been appointed Wheat Adviser to this Department and it will be his duty to see that the work in connection with at the different experimental farms is properly systematised and carried on in such a manner as to make improvements in the wheat industry.

He will supply the Farms with such seeds as he may deem fit, for the purpose of making varieties of wheat as are suitable to the districts of which the Farms are severally representative, and it will devolve upon him to recommend to farmers the best varieties to grow in their districts. He will also advise on the system of soil management in connection with wheat cropping on the farms, and is to have all facilities afforded to him to make experiments or test any points he may consider advisable with respect to any wheat crops that may be sown.

Mr Farrer will advise and make recommendations to the Department concerning all wheat work carried on at the Experimental Farms, and in regard to matters connected therewith. It has been intimated to him that he may look to the Managers of the various Experimental farms for any assistance he requires, and empowered to ask for it.

> *You are requested to be good enough to note the above and to afford to Mr Farrer every facility necessary tin carrying on his experiments.*
> *I have the honour to be, Sir,*
> *Your Obedient Servant,*

and it was signed by D. C. McLachlan.

These and later instructions reflect a lack of communication and oversight which were soon to be a matter of resentment. One possibility is that the matter of the letter was outlined to a secretary based on Campbell's initialled recommendation. In McLachlan's letter Farrer is classed as Wheat Adviser, not Wheat Experimentalist, and *will advise and make recommendations.* One flaw with Farrer and in his dealings with people his fiery temper, although curbed, was not far below the surface. However much the hindrance he suffered upset him he said that he could sometimes see the ridiculous side of things. W.S. Campbell was later to write,

> *He accepted the appointment with great reluctance, and would not have done so had I not urged him to join the department. Had I thought it possible that Farrer would be subjected to numerous obstructions in his work, to many difficulties raised, and to the humiliations suffered in the department, I would have advised him most strongly not to accept the appointment.*

At the end of his first departmental report Farrer wrote:

> *Since I have entered upon my office, and my work at Lambrigg has been carried on in the interests of Department, I have departed somewhat from my former policy in cross-breeding of aiming for results of a somewhat abstract character, such as increasing to an indefinite extent the flour strength and gluten content, and have rather directed my aims to making of improved varieties which are likely to be of immediate and practical value; and from this time, or in the immediate future, I shall hope every year to introduce into general cultivation through the experiment farms at least one or twwo varieties which are in some respects at least, improvement on the old standard sorts we have hitherto been growing.*

The series of letters in the Wagga Experiment Farm letter-book from Maurice McKeown to and about Farrer and what he was doing give ample evidence of the frustration on both sides. McKeown reacted strongly to what he saw as Farrer's arbitrary ways and lack of adherence to protocol and hierarchy. This turmoil was resolved with the establishment of the Cowra Experiment Farm for Farrer's experimental use in 1902.

The letters from McKeown to Head Office map out the path which the lack of communication took. McKeown saw the role of the Farm as being a 'show and tell' to farmers on how to practise their farming as a commercial operation. He saw Farrer's work as extravagant and of no practical use to the farmer. However, he was not the only one either then or later. In 1905 there was a controversy aired in *The Daily Telegraph* on the nutritional value of Sydney bread and the opinion was expressed that the Government was wasting its money paying Farrer to breed varieties which were not only valueless from the point of view of health but a confidence trick on the consumer.

In May 1898, before the appointment in August, McKeown wrote to Farrer, the private citizen:

I have the honor to acknowledge the receipt of your letter of the 9th and 11th instant and the various wheats forwarded by you. Please accept my best thanks for the trouble you have taken in this matter.

But almost from the beginning of the appointment, McKeown's letters tended to sound petulant and displaying a lack of understanding of Farrer's work:

8th November 1898:

In reply...I have the honor to state the course recommended by Mr Farrer had already been adopted by me and the varieties of field crops reduced by half. The varieties mentioned were sown by me as largely as possible with other popular kinds—72 acres of Australian Telavera are growing.

I advised Mr Farrer some time ago that steps would be taken to keep all varieties as nearly pure as possible by examination in the field.

Threshing is carried out with utmost care–the machines being cleared of every grain after treating each variety which is harvested and stacked entirely apart from any other.

14th November 1898:
Sir,

Referring to yours of the 3rd instant, I beg to state that I am keeping my crop records in a manner similar to that in which I kept them at Wollongbar, the following details being shown, viz: description, whence obtained, date of...?, previous crop, paddock number, area sown, quantity of seed, date of harvest, results and remarks. Records of rainfall and temperatures are also kept.

I don't think a more complete system could be adopted, but in addition to being shown on plans as suggested by Mr Farrer, it should always be kept in farm books, the plans being regarded as copies(?) of such book records.

One book should contain the records of the work of years, while if plans only are to be kept, the loss of a single plan could cause much inconvenience as, till it could be replaced, the chain of information would be incomplete.

12th April 1899 to Head Office:

...Mr Farrer asks for the employment of two extra men at 6/- per day each for a period of about two months in connection with wheat experiments. The method adopted by Mr Farrer is more costly than that hitherto practised. Mr Farrer objects to students labor and I cannot supply the necessary hands from regular Farm staff.

McKeown having stated his case, revised his options and wrote the letter of the 25th April 1899 to Head Office:

I have employed John Palmer and John Swaysland at wages of 6/- a day each to assist in the wheat experiments.

Later in the year on 19th October 1899 this letter was sent to Head Office:

Your memo of the 17th inst. duly to hand re experimental wheat plots. The proposed cropping of this 10 acres with cowpeas was in response to the written request from Mr Farrer which probably he has forgotten. However it is not too late to change plans.

Re the "befouling": of this land by horses it should be noted that the grazing by partly fed horses of which Mr Farrer complains was equivalent to one horse for 87 hours while it will take 4 horses five days to plough the land, therefore the dung of these horses which will be fully fed on oaten chaff will be liable to be dropped during 40 hours(or equal to one horse for 160 hours). When discussing the matter with him a few days ago I informed him that if the paddock was to be ploughed, it would be necessary to raise his objection to horses, to which he replied the dung of these horses would be of no consequence.

In July 1900 Farrer wrote Max Kahlbaum in Adelaide told him wryly that he was being a nuisance in the department.

The fact is I am in disgrace with the department, because I am such a disturbing factor. I want to see progress made and certain changes and they resent this. I fear my influence there at the moment is very small indeed. But I do not mind that.

By October 1900, McKeown himself was under some pressure. There was talk of closing the Farm as an educational facility as not enough students had enrolled because of lack of proper accommodation. It was only by the local Member of Parliament, James Gormly, persuading the Premier John See to visit the Farm, that matters were improved.

The following letter was sent from the Farm to Head Office that same month, with reference to wheat experimentalist's report.

I may say that I do not know what Mr Worboys reported but whether such report was favourable to me or not, I am satisfied that it would be accepted as that of a thoroughly practical man who made a full inspection of the crop under discussion. On the other hand Mr Farrer who is absolutely impractical, possesses physically defective eyesight, his mental vision is obscured by bitter personal feeling against those who differ from him and further he has "given his case away" by admitting he saw one side only of the crop and formed his conclusions from that point of view.

Mr Worboys and Mr Campbell inspected it from various positions and their reports are therefore more worthy of credence than the statements of mr Farrer.

An inspection at any time prove that Mr Farrer's report on blocks 2 and 3 which were sown for corn...? is utterly valueless and I strongly urge an early inspection by the Minister and yourself.

As has been explained to many enquires this block close to the road some years ago had burnt upon it some hundreds of cords of firewood and undoubtedly the soil has been greatly benefited by the deposit of ashes. In addition to this, part of the land to its whole depth was manured during Mr Valder's19 management. This extraordinary growth extends only a short distance into the paddock and represents far less than 10% of the block this year unmanured and on it Mr Farrer has based his erroneous conclusions heedless of the fact that a large area of inferior crop lay behind it. The results are in accordance with expectations as a substantial profit will be made on the application of these manures, both on the treated test block No2 and 3, each 25 acres showing a far heavier crop than No 1 untreated.

The board was removed to prevent persons of Mr Farrer's type receiving a wrong impression on seeing one position only and such impressions have always been dispelled on a complete inspection by genuine seekers after information.

Mr Farrer is in error in attributing the success of this crop to the previous feeding by sheep and it should be noted that on the patch admired and classed first by him a sheep was rarely if ever as they avoided this corner and all the other boundaries of the paddock owing to the constant traffic on the road.

As however it is alleged that Mr Farrer was to be seen on his hands and knees counting the number of dung deposits left by horses used in working the sheep he may be in a position to state exactly what quantity was left there by the sheep.

My experience with Mr Farrer prevents me from regarding him seriously as an authority, my first shock having been a recommendation to apply manure costing £4 an acre. Later, when his views as to manuring were modified as to quantity and cost he seemed unable to form any definite ideas as to his requirements, frequently changing his requisitions. On one occasion he ordered an orchard manure for wheat then charged me with the error but I had his instructions in writing.

... He has been instructed by you to confine his attention to his own area

*and to refrain from making suggestions till he is asked for them and I respect-
fully request that his position again be stated to him.*

The misunderstandings went on. That same lack of comprehension of scientific experiment was not only within the New South Wales Department but was
to be found in other States. In Victoria, Howard Pye, a friend and fellow wheat
breeder of Farrer's for many years wrote to Farrer in 1904:

*Personally I have been absolutely astounded that your government did not find
your worth years before it did and in America what you do would have been sent
throughout the length and breadth of the country, and you would have received
the salary of the first order. Here(Victoria) we have men in public positions who
have never been at Dookie or with you and in rosy language tell of the wonders
of crossbreeding of wheats done in other countries and ask, 'why is it not done
here?'*

Pye like Farrer had limited skilled labour and lack of funds.

In 1900, with the release of the two varieties Federation and Bobs, and his
international standing Farrer could be judged to have proved his worth, but it
was only with the establishment of the Cowra Experiment Farm in 1902 specifically for his wheat experiments under the management of G. L. Sutton that the
pressure was relaxed.

PROBLEMS ARISING

ALTHOUGH TRADITIONALLY in the old world farmers had the experience of thousands of years of farming practice to maintain productivity on their farms, this was rarely the case in Australia. The poor management of the productivity of their farms was compounded by the unfamiliarity of the soil and climate. In New South Wales there was little knowledge among the people most affected by the fertility of the soil which they utilised. From 1892 when F.B. Guthrie had first joined the Department of Agriculture part of his brief was to help farmers were seeking some idea of the quality of their soils. The haphazard nature of samples submitted and exigencies of the time and space available, prompted Guthrie to write a piece in early 1898 in the *Agricultural Gazette*. He called it 'remarks' on the object and method of soil analysis:

> *No information of any of these essential points is afforded by examining a few ounces of soil dumped down in front of one without a word...*
>
> *Much more advantage might be taken of this fact than is at present done. If a number of those interested in farming were to found social clubs or Farmers unions in different districts where they could discuss topics of common interest it would be found that a number of questions could be referred with advantage to the Department, and the answers communicated in this way to a larger number than is now the case.*

Soil and its fertility was of more than academic interest. In the world at the end of the 19th century it was becoming a matter about which politicians were being made aware. In 1916 Sir William Crookes edited a book, *The Wheat Problem* in which he and others addressed problem as it appeared politically. Included was not only his 1898 address but commentary on it written for the book. In his 1898 Presidential Address to the British Association for the Advancement of Science, Crookes had expounded on the approaching starvation facing the people of the world because of the declining fertility of the soils of the wheat-growing countries in the temperate zones of the world. He surveyed the capacity of the wheat-growing countries to produce wheat, but the rate at which the wheat-growing land was being taken up for growing the grain, would lead to the last available acre being under crop. Then nitrogenous fertilisers must be used to raise the yield of each ear of wheat, but, *to do this efficiently will exhaust all the available store of nitrate of soda. For years past we have been spending fixed nitrogen a culpably extravagant rate, heedless of the fact that it is fixed with extreme slowness and difficulty, while its liberation in the free state takes place always with rapidity, and sometimes with explosive violence.*

He laid before his audience the existing sources of nitrogenous fertiliser: sulphate of ammonia from the gas-making process; the ability of leguminous plants to fix nitrogen; the nitrogen in the drainage and sewage from towns. By his calculation none of these would provide enough for the increase of yield and the nitrates brought out from places like Chile were all finite.

Let us remember the plant creates nothing; there is nothing in the bread which is not absorbed from the soil, and unless the abstracted nitrogen is returned to the soil, its fertility must ultimately be exhausted. When we apply to the land nitrate of soda, sulphate of ammonia, or guano, we are drawing on the earth's capital and our drafts will not perpetually be honoured.

He, had however, a chemist's solution to the problem in one specific source, the harnessing of available nitrogen into fertilisers. He proposed that efforts should be made to achieve the fixation of atmospheric nitrogen by electricity, regarding it as one of the greatest discoveries awaiting the chemists. *The fixation of nitrogen is a question of the not far-distant future. Unless we can class it among certainties to come, the great Caucasian race will ceased to be foremost in the world and will be squeezed out of existence by races to whom wheaten bread is not the staff of life.*

William Farrer was pursuing another approach to the problem by developing new varieties suitable for expanding the boundaries of the wheat-growing areas. However, he was always aware that his observations and theories of climate and soil needed to be proved scientifically. So it was that almost thirty years after Farrer had published his paper, *Grass and sheep-farming* with its analysis of soil and its requirements, he returned to the matter of soils. Farrer read a paper at the Australasian Association for the Advancement of Science in 1901, giving it the grand title, *The absolute dependence of Agricultural Progress upon Experiments and Suggestions in regard to some Directions in which Experimental Work should be done for Agriculture in Australia.* The capital letters announce the vehemence he felt about the whole topic of experimentation considering his own experiences with the determination of people in authority to resist change or improvement, always casting a jaundiced eye on any experiment which was too expensive in both money and time.

He set out the need for work on the fertility, leading to increased productivity, especially in the semi-arid and arid parts of the country. His experiences and work in the Western Division always lent weight to what he was trying to achieve for New South Wales in the matter of wheat-breeding. The conditions about which he was concerned were, *in those parts of Australia where remoteness from the commercial centres on the coast, the high freight charges of the railways, the dryness of the soil, and the uncertainty, whether there will be moisture enough to make chemical manures available for the crops they are applied to, make artificial fertilisers to be used at great disadvantage.* His wider world view: *...of almost, if not quite equal importance, that the improvement of Agriculture in the direction I am indicating should be taken in hand in the same manner the world over.*

In his pamphlet of 1873 he discussed the chemical constituents of the soil, and now said that, in general, there was ignorance about the nature of the soil's elements and how they were combined and associated. His remarks reflect discussions with Guthrie on the work he had been doing with soil and artificial manures. Farrer maintained that the more complex the constitution of the soil the more the chemical equilibrium was disturbed by other activities, the greater would be the soils ability to provide plant food. The chemical activities of the soil were not the only forces at work: bacteria contributed to the breakdown of organic matter to produce nitrogenous food for plants. This was in contrast to Crookes, who expressed the view that this organic matter was a dangerous infectious material.

For Farrer the availability of nitrogen sources in the soil had been an abiding interest from the 1880s when he had begun to plant crimson clover, *trifolium incarnatum*. He had often from that time written numerous articles on its value for the farmer. While he had made observations about its value, it was on matters such as this that he was seeking experimentation to have proven results rather than anecdotal. He was calling for improvements in the methods of managing the soil to enhance its capacity to increase yield. Although he had his own ideas on the physical condition of the soil: the use of organic matter; the increase in chemical activity; subsoiling and the right plough to use, properly constructed experiments should be carried out to discover the best methods for the best results. In particular, he wanted this to apply to increasing and maintaining fertility in the drier part of the country. Theories could be discussed as he himself did, but as always with him facts were what was needed.

Guthrie became President of the Royal Society in 1903, addressed the matter of co-ordination of scientific work. He made an overview of all the separate departments and institutions which employed analysts of everything from mineral to kerosene to soap maintaining separate laboratories to carry out the work. In his view this did not make for greater efficiency but removed any possibility of meaningful *investigation of subjects of importance to the community* because of the constraints imposed by routine work and administration imperatives. His own departmental experiences had strengthened his opinion on the matter. Decentralisation was a wasteful exercise.

Crookes' argument and Farrer's proposal for experimentation to address the matter of productivity would require additional disciplines to be brought to bear, needing Guthrie's concept of an overarching organisation. In the new Commonwealth, under Guthrie's and others impetus The Commonwealth Advisory Council of Science and Industry was set up in 1916 almost as he had outlined it in 1903.

EXPERIMENT FARMS 1900 AND ON

NATHAN COBB IN HIS report of 1891 had argued that experimental land was necessary for the Departmental officers to do research and carry out investigations. The Department of Mines and Agriculture from the early 1890s was forced by the financial situation of the time as well as the public calls for the education for the practice of agriculture to combine experimental work with the education in agricultural pursuits. When the Murrumbidgee Model Farm as it was first called and the orchard was being established and the arrival of students imminent, Benson said that the students would be the workers for which they would be paying a fee of £25 a year. Their practical instruction would be the source of manpower. It was originally intended that the experimental crops should be horticultural but the exigencies of the wheat problem, Cobbs interest and involvement and the need for land for experiment, having regard to climate and soil typical of the area, saw Wagga Experiment Farm designated for the purpose of experimenting with wheat. But as yet there were no students at the farm.

In 1894–95 with the inauguration of Government experimental farms the Minister for Mines and Agriculture, Sydney Smith put an emphasis on fodder and feeding experiments He laid out how the experiment farms should operate:. *In regard to the study of different grasses…exact experiments instituted, similar to those adopted in wheat experiment stations…*He specified the kinds of experiments to be done in manures, they *will be of two kinds, field experiments(subdivided into permanent having an educational object and experimental plots for the purpose of testing different conditions) and more exact pot experiments.* all the educational experiment plots where the results were already known would be arranged so that they could be readily seen and understood by both farmers and students. He laid out the intended use of the experimental plots: on them there were to be experiments to explore the most favourable methods; the most suitable manures; the varieties of seed as well as the applicable time and manner of planting. The pot experiments would be similar to the field experiments but they would under controlled conditions, not affected by differences in soil and climate. While the Minister's reported directive may have been of good intent, the financial constraints and the lack of enough trained personnel would prove to be a constraint.

Nathan Cobb, however, was to set a departmental example for experimental work. He was able to persuade the Minister, Sydney Smith through the good offices of W. S. Campbell perhaps, to have a laboratory built at Wagga Experiment Farm. In his report for 1897 he spoke optimistically of experiment farms saying that, *they were no longer an experiment in this country. They have justly won popular favour and have come to stay.* His enthusiasm, however, was not shared by all the members of the establishment. Stewart Mowle a Member of the Legislative Council writing in a letter to Farrer in 1898, expressed the view that government farms were just a fad.

In November 1898 the Minister of Mines and Agriculture paid a quiet visit to the Farm to see for himself how it was progressing. It was not an altogether felicitous time to come but he could see for himself the unfavourable results of the severe drought. He was impressed with the canning and drying and agreed that it was necessary to erect a more suitable building for the cannery operations. It would cost an estimated £2000. He also agreed to have new tank excavated below the wheat crops to the west of the hill. The pressing questions of new quarters for the students was brought to his attention but he thought it was improbable that *any steps in that direction will be taken for the present.* This marking time on improved accommodation was to have a negative effect on student enrolment at the Farm by 1900. This in itself created and economic problem for the viability of the farm, when the cost of running the Farm was offset by student labour.

Maurice McKeown was, like George Valder, one of the early officers of the new Department. He came to the Wagga Experiment Farm from Wollongbar where he had carved out that Experiment Farm from the North Coast rainforest. He arrived at Wagga Wagga towards the end of 1897 taking over from Nathan Cobb who had been acting manager after Valder went to Hawkesbury. With his arrival in Wagga Wagga there was a change of tone in the letters in the letter-book. In fact the correspondence in the letter-books reveal the personality of the man far more than any of his somewhat pedestrian pieces in the *Agricultural Gazette,* quite unlike Cobb. He was a man who wrote (and spoke) his mind and kept nothing back whether to the Under Secretary, parents or students. His first letters were reports sent out to the parents about their sons' results and conduct. They give an inkling of the direction he intended to take: *Student Shaw, 5th place. His conduct has not been satisfactory, chiefly in the direction of being tardy in getting to work.*

James Gormly the local Member of Parliament made representations for the Premier to visit the Farm on his way back to Sydney from a meeting in Albury. He said that the Farm, in several respects, was not what it ought to be. He pointed out that it was not sufficiently attractive for students, and not popular enough to be of interest to farmers and the general public. The Premier along with a number of his ministers, made the visit and recognised that what Gormly had said was a matter of fact.

The Premier in an interview with *The Wagga Wagga Advertiser* said:

Recognising the importance of technical education especially of such a character as that to be afforded on a well-conducted Experiment Farm, I am most desirous that the Wagga Farm should be made as attractive as possible from a practical point of view to students, farmers and others interested

He went on to stress the need for *the rising generation to choose industrial occupations other than follow the fashion of being unduly influenced by the exaggerated importance of scholastic distinction.*

The concerns voiced by the newspapers were that education for farmers was an important issue stressing that it should be practical. The experimental side of the Farms was to be considered part of this education and less concerned with science. Nathan Cobb when he had been acting manager in 1897 had obtained the scientific equipment of science such as microscopes for the students. Guthrie, in 1898, outlining his ideas on agricultural education, stressing the need for the students to be trained in scientific method which developed habits of observation and careful reasoning. Student education on the experiment farms then was given after the work they had done in the field as practical training for their apprenticeship as farmers. Scientifically trained tutors would be the visiting experts.

When Farrer joined the department, McKeown was not pleased when Farrer refused to have students on his plots where everything was so precise but he did have a succession on individual young men staying at Lambrigg learning his methods; Jim Pridham for instance, who succeeded George Sutton at Cowra, and George Norris, who worked later with Guthrie on wheat and flour investigations. For Farrer, Experiment Farms should be for the purpose of experiment and trial not be for general farm training.

For a number of years before 1898, Farrer had been conducting his own wheat experiments on sites other than Lambrigg. With departmental permission these sites were on the Experiment Farms at Hawkesbury, and Bathurst. After 1898 and his appointment, it was his practice to visit each of the Experiment Farms in the spring where he would inspect the plots where the crossbreds which he had grown at Lambrigg were being tested and the new varieties fixed. He would make an assessment of what was happening and then return to Lambrigg in the early summer to continue his own breeding programme, leaving instructions about what should be carried forward. McKeown, because of financial constraints found this habit too much for his management style. He wrote on the 25th April, after Farrer had requested records from Bob Hurst, the assistant on the wheat plots, which McKeown said were official documents: *I have to inform you that unless your instructions are forwarded through the Manager of the Farm no notice will be taken of them in the future*

Relations between Farrer and McKeown continued to be frosty, especially when Farrer wrote to the Under-Secretary complaining that McKeown was obstructing the experiments by not giving him the labour. In return McKeown wrote asking for Farrer to be directed to give particulars of McKeown's refusal to comply with the wheat-breeders requests. A later letter from McKeown to the Under-Secretary said:

The case stands thus— I have assigned a man Hurst to Mr Farrer and the latter each year applies for additional costly assistance which notwithstanding Mr

> *Farrer's assertions is paid for by the Farm…*
> *…I submit that Mr Farrer having an imperfect grasp of the expense and labour involved, lays out more work than can be properly carried out.*

The controversy went on for two years, until Farrer had cemented his reputation as a wheat breeder with the rust-resistant wheat, 'Federation'.

McKeown was not alone among the farm managers in dismissing Farrer as a crank, a fact which Campbell himself recognised.

1900 was the last year of the 19th century and the passing of the British Act of Parliament to allow Australia to have a Commonwealth of Australia Constitution. It also the year when Australian volunteers were involved in wars which, were the last under British command; regiments were raised by individual men of wealth to go to fight the Boers in South Africa and later in the year the last contingents of the separate colonial navies of New South Wales and Victoria sailed for China to be involved in the matter of the Boxer Rebellion.

In 1900 Farrer, while he was making demands for improvements and making himself unpopular in the department. Early in the year Farrer wrote in the *Agricultural Gazette*:

> *The efforts which were begun some ten or twelve years ago in Australia to help the agriculture of the country have lead to the establishment of experiment farms in most of the colonies. It seems, however, to be doubtful if at the time of their creation more than hazy ideas existed in regard to the manner in which these farms were to do the good work which was expected of them: and, indeed, it is far from certain even now whether much progress has been made in thinking out the matter.*

Later that same year Farrer was asked to deliver a paper to the Congress of the Australasian Association for the Advancement of Science in Melbourne that year, making a case for *How Experiment farms can be made to help in the best manner the Agriculture of the Country.*

People such as Farrer were regarded as *scientific gardeners* and in his address he remarked that *agriculture was an experimental science or an art.* He drew attention to the experiments continuing at Rothamstead connected with tillage, methods of seeding, and manuring which were carried out in a systematic manner: care was taken with reports of the results as well as the various factors which affected them. In Australia with the exception of progressive farmers, the occupation of agriculture was being conducted along traditional lines, which had either been brought with them or learned from their fathers, ignoring the changes in the antipodean soils and climate. He said, *sufficient time to unlearn the practices have barely elapsed,* and the lack of the guidance which experiment farms could provide could be felt in all branches of agriculture. The would-be progressive farmers were driven back on their own experiments which were often unrecorded and unsystematic. Not only that, but they were often kept secret: not subject to criticism or examination. They were a cost to the farming community because they were costly, not only in time and energy, but money. It was better if the same time, energy and money were put into government experiment where even the failures could be of value. Then the approach would systematic and evaluated. He was to say in 1901, *The only plan which remains is the old one of actual experimental work–of trial and error–we must make the best use we can of our knowledge of principles for planning our experiments and when they fail, for finding out the causes of our failures, which we must go on eliminating until we have succeeded.*

He like Cobb was not afraid to speak out about what he thought was wrong and needed to be improved. Cobb had not yet returned from his tour as Special Commissioner investigating scientific facilities in Europe and America and making connections for the future. Guthrie too was being outspoken. In the *Agricultural Gazette* of 1900 he wrote an article, *Chemistry in Agriculture:*

> *…It is sufficient to say that the science of agriculture progressing along strictly scientific lines and engaging men of all countries of the highest scientific attainments…results will be obtained which will entirely revolutionise our present methods of farming.*

Guthrie was at the time making a point in answer to the practical men who were impatient of delay and wanted results which would fill the wheat bags for a good price, not wanting to take into account the time it took to produce a new variety which would suit the requirements of the industry. It was under these circumstances that Farrer released Federation in 1901.

It was this year that the Government of India sent a high- ranking officer, Mr Moreland, to New South Wales to consult with Farrer on his methods. Farrer's reputation as a world leader in wheat breeding had been established especially in Canada and the United States. W.J. Spillman, in a paper read on *Quantitative Studies of the Transmission of Parental Characteristics by Hybrid Offspring* to the 15th Congress of the Association of Agricultural Colleges and Experiment Stations in 1901, declared concerning hybridisation, that, *until recently practically all the work was done with ornamental plants only or with those of little economic importance. More recently much has been done with field crops.* Linking Farrer with European and American names such as Rimpau of Germany and Carleton of the US, he continued, *he has attracted general attention to the possibilities of improvement by hybridisation and selection.*

There is no doubt that with this reputation and with the success of the release of Bobs, Comeback and Federation that same year, Farrer could mount a case for Experiment Farms without students. Coolabah Experiment Farm had been established in 1898 at the request of the then Minister for Lands, Joseph Carruthers, who was opening up land to closer settlement and wanted farmers settling in the West Bogan; Glen Innes was established in 1902 as being the most representative climatic district in that of the State. Different climatic districts served for Farrer to use for the testing the wheat he had bred for the variations in climate.

What Farrer wanted was an Experiment Farm run on his lines geared to specific purpose of wheat breeding and the production of pure seed. Cowra was the result.

COWRA AND GEORGE SUTTON

G EORGE LOWE SUTTON was perhaps the one who was most in tune with Farrer's ideas about wheat breeding and most prepared to follow Farrer's philosophy and example. He was one of Farrer's 'apprentices' unlike Guthrie whom Farrer would have regarded as an intellectual equal

Sutton as an experimentalist worked closely with Farrer and when he died carried on his work in New South Wales and when Sutton went to Western Australia as Wheat Commissioner he took with him the philosophy and the ideas about wheat breeding which he had absorbed from Farrer, to the advantage of the West. For Farrer he was a trusted confidant, who, from the 1920s on helped to preserve Farrer's memory.

He was born in 1872 in England at Liverpool where his father was a shipping agent but who died when the boy was six months old. In 1882, his mother, Ellen, brought him to New South Wales, to live in Sydney where George went first to Fort Street Model School and then on a scholarship to Sydney Boys High School. After he left school he went to work on his uncle's dairy farm near Liverpool, now a suburb of Sydney, but then a country town. No doubt, being a bright boy and showing an interest in agriculture he began study at Sydney Technical College, distinguishing himself by being awarded a Diploma of Agriculture with first class honours.

The biographical note in *The Cyclopaedia of Western Australia* by J.S. Battye, says that after about two years on the dairy farm, he took a trip *among the pastoral areas on the borders of Queensland in order to enlarge his experience.* Taking a year, he then returned to Liverpool and dairy-farming, which he gave up when he was appointed as experimentalist at Hawkesbury Agricultural College. In 1900 he became lecturer in agronomy and experimentalist.

From 1893 Farrer had been able to experiment with 65 varieties on the Hawkesbury Agricultural College Farm or perhaps on about $1\frac{1}{2}$ acres which is recorded as having been used in 1899 when 400 strains were grown. In 1905 Farrer said that the experiments which were carried out at the College were mainly for the purpose of testing rust-resistance of the varieties which might prove suitable for coastal regions.

At Hawkesbury Sutton was conducting maize experiments side by side with Farrer's wheat trials. As he supervised the wheat work, Sutton found that the methods Farrer was using would enhance his own work with the maize. Sutton introduced a system whereby experiments could be controlled: experiments are replicated and tested against a control. The design he initiated was the 'three plot system' where one plot was the centre of a group of three which acted as the control. As in the case of his maize, so of any wheat experiment the centre plot was the standard variety in any trial. His experiences at Hawkesbury were the basis of the work he would do at Cowra and later promote in Western Australia.

While Sutton would have been accepted by most members of the (sub)Department of Agriculture as a 'practical man', he was able to appreciate the value and potential of Farrer's work. It was not only from a working standpoint but on a personal level. He wrote to a Farrer connection in about 1937, *he was my closest friend.*

He and Farrer were not unlike in temperament. Eric Lawson in a biography of George Sutton in *'Westralian Portraits'*, wrote that *he had little time for those he considered less than dedicated and did not readily forgive those who crossed him professionally.*

People keep letters reminding them of important points and people in their lives. George Lowe Sutton kept in his papers six or seven letters from William Farrer belonging to a period between 1903 and 1906, the last being written a few days before William Farrer died.

W.S. Campbell, prompted by Farrer's request for Sutton to be appointed manager at Cowra, wrote to Sutton in October 1903,

> *Dear Sir*
>
> *I had an interview with Minister respecting your application for an increase and suggested to him that you should be appointed to the new Cowra farm as well as to supervise the Experimental work at the college and Mr Farrer's experiments on the various farms.*
>
> *Mr Kidd is quite favourable to my proposition and advises that your application for an increase be held over until we can settle up matters which will done as speedily as possible.*
>
> *Kindly keep this to yourself for it will be better not to make public anything that may imperil the carrying out of the arrangement.*
>
> *I think that the new billet will be an immense improvement on your present one and you will have a big field for future work.*

That Campbell and Farrer had discussed the proposal informally, perhaps without Sutton's knowledge, may be inferred from the letter Farrer wrote him on December 25th 1903:

> *…a better one than would be the position of the head of an agricultural college anywhere and especially of one in what cannot be but for a long time be a relatively stagnant one like W. A. I do not think you would like W. A. at all on that account; I was at Albany for a few hours several years ago and I felt the atmosphere at once: also I had as fellow passengers several men from that State, some of them in high positions in it and I could not help noting the inferiority of these men to those who occupied corresponding positions in my own State. If*

*you want to be in the forefront of progress in Australia, you will have to remain
in an Eastern State, where we want you to remain so much. I think that your
prospects in this State would be better than they would in W. A. even if you
appeared to go ahead more quickly there at first: and if a Federal Department
of Agriculture were created, and if at the time of its establishment you were
found to be carrying on systematic experiments and had had experience of such
work, it is more than likely that you might be asked to take a permanent position in it.*

Farrer's forecast proved to be correct, as in 1911 Sutton, having accumulat-
ed the experience of different soils and climates with his management of Cowra
and Coolabah after Farrer's death, was head-hunted by James Mitchell, the
Western Australian Minister for Agriculture to Commissioner for the
Wheatlands.

That Sutton was still without any firm commitment from the Director is evident from the following letter Farrer wrote on the 30th of December

My dear Mr Sutton,

*I was very glad indeed to get your letter today. I think the very best thing you
can do is to see Mr C. I have just written to him and told him that if you came
up here to help me, you would make it possible for me to get in terms with the
work I have to do preparing for the sowing of the experimental plots at the different farms. I have told him that the late character of the season has put my
work back, and that the rains we are getting almost daily is causing my progress
with the harvesting to be very slow indeed. I have told him that by coming up
here for a month or so, you would get into touch with my work and get a knowledge of it, that you could look after the wheat experimental work which is going
on at the farms when ever you went to them. I also hinted that the present opportunity was one you might never get again and that you were unwilling to let it
slip away unless you had cps something tangible here. Go down and see Mr C.
and press matters from this point of view.*

*I do not think you need trouble yourself and think that you are quite safe,
but I think you ought to do what you can do to hurry on things.*

*I got a bit of a surprise yesterday which I have not yet recovered from and I
still have to digest the news. You know that last year I sent the Department of
Agriculture in Melbourne 2 bushels of 'Bobs' from Bathurst and 2 bushels from
Wagga which I purchased from these farms for10/- a bushel, to be distributed in
place of 'Rerraf' which had been distributed before its inferiority in the mill had
been ascertained. I daresay you recollect the letter I wrote to The Australasian on
the matter. Well, the seed from Bathurst was all right for I planted some taken
from the bulk, and detected no strangers in the drill: but a correspondent who
got some of the Wagga seed told me it was so mixed with a rusty lot so that the
crop is useless for seed purposes I am preparing to write to the Ag Department of*

Victoria asking them for what they promised to send me— a report on the behaviour of 'Bobs' in that State. I shall also ask them to tell me whether the supplies I sent them were pure or not and whether there was any difference in this respect between the two lots. I did not send an order to the Wagga Farm until after my letter appeared in The Australian so that McK must have known that it was my most important that the seed should be pure. Of course my opinion is that the seed which they sent was made to be impure but it would be very difficult to get to the bottom of the matter.

I am glad to hear that you are pollinating pumpkins. The present rain are very fast bringing up the seed you gave me and I fear our plants will be too late to do any good.

Tomorrow I have to make an early start for Queanbeyan. Mr Allen is to be up and I want to meet him and see about the mending of a smashed buggy.

Your experiments with the Triangular pumpkins is an interesting one.

A report on the bunt experiments came from Bathurst yesterday and it appears that a few of them which were entirely free from bunt last year and are quite clean again this year. They are making me feel more hopeful.

With kind regards and a happy New Year.

Campbell later expressed his opinion on the matter of Cowra,

The establishment of a small experimental farm at Cowra to be worked shortly in conjunction with the Coolabah Farm under Mr Sutton will be of much benefit to Mr Farrer. It is proposed to limit the work of these two small farms almost entirely to experiments with wheats. The experiments at Coolabah should prove extremely valuable for, although outside the limits of profitable wheat production, I believe that carefully conducted and well thought-out experiments on that Farm with wheats made by Mr Farrer or varieties obtained from other countries and recommended by him, the present wheat-growing areas of the State will be considerably extended to include a vast area of land suitable for wheat as regards soil, but at present considered unsuitable on account of insufficient rainfall.

These letters had been kept by Sutton, no doubt recalling for him the advice and assistance Farrer afforded him in the progress of his appointment to Cowra. In September 1904 Farrer had written a formal letter to the Director which set out Farrer's view of the purpose of the Cowra Farm. The copy is in the Sutton papers and by inference lays out the impediments which Farrer found in his dealings with the other experiment farms.

Cowra Experiment Farm was an initiative which Farrer had taken up with Campbell, perhaps early in 1903, and a letter marked 'Non-Official' from Campbell to George Sutton in October 1903 indicates the shape of Campbell's advocacy to the Minister, Mr Kidd. Farrer having found a man after his own heart whose youth would be able to take on what was becoming an increasing burden to the older man.

Dear Sir,

I had an interview with the Minister respecting your application for an increase and suggested to him that you should be appointed to the new Cowra Farm as well as supervise the experimental work at the College and mr Farrer's experiments on the various farms.

Mr Kidd is quite favourable to my proposition and advises that your application for an increase be held over until we can settle up the matter which will be done as speedily as possible.

Kindly keep this to yourself for it will be better not to make public anything that may imperil the carrying out of the arrangements.

I think the new billet will be an immense improvement on your present one and you will have a big field for future work
Yours truly
Walter S. Campbell

The implications in the letter were that there were some elements in the Department, and most probably outside, who were unlikely to be favourable of any concessions to Farrer or what they saw as favourable treatment of Sutton who be side-stepping the normal steps of advancement.

By Christmas time Sutton does not appear to have heard anything on the matter and was feeling that nothing would come of it and must have written to Farrer about his misgivings of the whole proposal. Farrer persuasively refuted Sutton's misgivings in the letter dated December 25th

Mr Dear Mr Sutton,

You are not being humbugged, that I can answer for, for in a note I received yesterday from Mr C. and he says briefly 'I am busy about Sutton and I think can arrange matters. I saw the M. about it and he is quite ready to appoint Sutton to Cowra etc'.

He continued his letter,

…You speak of your want of knowledge and insufficient education for the work. These are details which are easily removed and the knowledge which is gained because it is badly wanted is always more thorough and better than the knowledge which teachers impart to you in the course of a general education. I too am very ignorant of many things we shall want—Mr Guthrie would impress that fact on you—but I am not afraid for all that. I can study things and seize upon the points I want and go to work with them. Although so old I feel that I am still a learner.

…I shall urge Mr C to have your appointment made quickly and shall ask that you be set to help me with the wheats and the threshing of them, in order that we may talk matters over and mature planning: and you can get a better insight into my work

> *I was too tired yesterday to give much attention to the experiments you dealt
> with in your letter: but I will look at them today and try to take them in. My
> mind is kept occupied by my wheats which have been badly beaten down by the
> rains.*

Farrer was presenting a persuasive position to Sutton, appealing to the
advantages to be gained if he stayed and was not above applying a little emo-
tional blackmail. He was not prepared to let the matter drop but was prepared
to be insistent.

The interval between one letter and the other saw Sutton plagued with
doubts and impatient for a result. He wrote again to Farrer who replied on the
30th December encouraging him to be pro-active.

> *My Dear Mr Sutton*
>
> *I was very glad to receive your letter today. I think the very best thing you
> can do is to see Mr C. I have just written to him and told him that if you come
> up here to help me, you would make it possible for me to get on terms with the
> work I have to do preparing for sowing the experiments at the different farms. I
> have told him that the late character of the season has put my work back and
> that the rains we are getting almost daily is causing my progress with the har-
> vesting to be very slow indeed. I have told him that by coming up here for a
> month or so, you would get in touch with my work and get with a knowledge of
> it, that you could look after the wheat experimental work which is going on at
> the F when ever you wen to them. I also hinted that the present opportunity was
> one that you might never get again and that you were unwilling to let it slip
> away unless you had something tangible here. Go down and see Mr C. and press
> matters from this point of view.*
>
> *…I do not think that you need trouble yourself and think that you are quite
> safe, but I think you ought to do what you can to hurry on things.*

The rest of the letter dealt with other matters.

Sutton was appointed– but Cowra began in a similar manner to Wagga
Experiment Farm– with no proper accommodation for the new manager as a
married man. Farrer again had to encourage him to raise his spirits. It was not
however, until April 1905 that the Farm was officially opened. when it was reported that 200 of the 990 acres of the new Farm had been cleared and the first crop
was about to go in and it was noted approvingly that the experiments would *aim
at production at minimum cost.* No money was to spent on the Manager's house
which was *of primitive quality only:* no improvement in accommodation was to be
expected until finances improved. The main thing for Farrer and Sutton was that
they had their new Experiment farm which was to be for experiments only.

Among George Sutton's saved correspondence is a letter which Farrer
wrote to the Director of Agriculture on September 23rd 1904 about six months

after the Cowra Farm came into operation. Farrer never one to stand back if something needed to be said, and although the Farm was not specifically under his control, it was designed to fulfil his needs in experimentation.

Sir,

In connection with the very great reductions, which are being made in the staffs of the Experimental Farms, I have the honor to bring before you the case of the Cowra farm, in which it is proposed to give the manager the assistance of only one man. I have always been given to understand that this Farm is to be what the older Farms, such as those at Wagga and Bathurst have never been, an Experimental farm and devoted specially, if not entirely, to experiments connected with the wheat-growing industry, and that in it with the manager, Mr G. L. Sutton, be enabled to carry on experiments for the purpose of working out improvements in the methods of it I would, in association carrying on that industry in our climate. If this were done and if only a moderate amount of success attended our efforts, it would bring far greater benefit to the country than have or ever will those farms which are run for the vain purpose of making profits. I have the honor to point out that in the practices of successful wheat-growing, in common with those of all farming, are founded on experience, and that as such additional experience as is needed to improve our present practice and cause it to more suitable for our conditions, can only be acquired by means of systematic and carefully designed experiments, the wisdom of making experiments for the purpose is obvious. Although I think it would not be wise to enter on experimental work of this character at once on a large scale, and that it would be better to begin quietly and allow new operations to expand naturally, I believe that with only one man at Cowra we can do nothing of any value. I think, however, that with two men we might make a modest beginning. The services of one man, indeed, will be required for wheat half the year, for the experiments I have arranged to carry on at Cowra in connection with making bunt-resisting varieties of wheat–work in which I have already met with very satisfactory success. I may mention also that I have lately measured out a number of plots for experimental purposes on the Cowra Farm.

As I am told that the Coolabah Farm is being kept going mainly for the purpose of enabling me to discover or make wheats which are specially suitable for our dry interior, and as, although I know that such wheats can be made, I know that they would advantage us very little unless a knowledge were gained of the modifications of the industry processes of cultivation which are demanded by the dry conditions of our interior., I have the honor to suggest that the Coolabah Farm be, like the Cowra Farm, devoted entirely to experiments and joined to that Farm by being placed with it under Mr. Sutton's charge. If this were done, we could carry on our experiments to much greater advantage, and have them under the closest possible control;: and this is absolutely necessary for a correct comparison of results to be made.

I have the honor to be Sir
William Farrer.

The opening up of more land for settlement from the 1880s, the mechanisation of wheat farming, gradual expansion of railways into the hinterland, brought about an urgent demand for drought-resisting wheat varieties for the lower rainfall areas. Farrer from his surveying days had become aware of the potential of the western plains with the heat and the low rainfall... So that his drought-resisters could be tested in a typical area, trials were made at Wagga Experiment Farm. In 1898 Coolabah Experiment Farm, in the Bogan Scrub, between Nyngan and Bourke became a site for the trials different methods of farming wheat with different varieties of the grain. There were experiments with other grains, grasses and vegetables. Only the stations were likely to have green vegetables and the new farmers needed to know the methods of cultivation, not being likely to have the benefit of a Chinese gardener.

Farrer's practice was to visit each one of the Experiment Farms four or five times in the year. Although a manager was installed, before Cowra Farm was dedicated and Sutton given the supervisory role, Farrer himself went there in his regular rounds of the experiment farms to assess the progress of his varieties in that soil and climate

In time, because of the low fertility of the soil allied to the harshness of the climate and the likelihood of heavy frosts, the whole operation was moved lock stock and barrel to Nyngan in 1910, on George Sutton's advice as the managing of both Cowra and Coolabah.

When George Sutton years later wrote a history of the Western Australian wheat industry he recounted his experiences with farmers and the bunt problem.

Dr Darnell Smith(biologist N. S. W. had been co-operating with Mr Farrer and myself in the bunt research work carried out at the Cowra Experiment farm and had proposed copper carbonate as likely to prove equal to and more convenient than standard bluestone and limewater treatment of the prevention of bunt. At the same time it was claimed to be effective against re-infection. His tests proved his contentions to be sound and the copper carbonate dust treatment gradually replaced the old wet treatment.

Farmers believed that 'bunt' was a *visitation from God* and therefore uncontrollable.

One of Farrer's ideas which Sutton took with him to Western Australia as Commissioner for the wheatlands was to have experiment farms where pure seed from which could be supplied to farmers. Because there were not the experienced workmen to undertake the work it was they were not used to the care which needed to be taken and that many farmers.

FARRER AND KAHLBAUM

ARRER, FROM ABOUT 1893, had subscribed to *Garden and Farm* a magazine published in South Australia, approximating the *Agricultural Gazette in New South Wales*. In 1896 he came upon a piece under a nom-de-plume, 'Mylegs'. He thereupon wrote to W.C. Grasby at Roseworthy Agricultural College whom he had known in the past and asked for a contact with 'Mylegs'. He explained his interest:

> *I for one am much obliged to the writer who is evidently a miller, for giving us the definite information he has given. To give any information at all is what the millers have hitherto refused to do.*
>
> *I would like to learn if Mylegs has no objection to letting me know by which means he ascertained the wet and dry gluten. If by means of the 'Aleurometer' as he mentions that he had only a few ounces of grain of the wheats he examined. I suppose that his examination was made by means of a small roller mill–such as are now made for the purpose of testing the milling value of samples of wheat.*

This particular letter was attached to the collection of letters in the Farrer Collection which had been sent to Max Kahlbaum, a miller, who had had what Farrer called *a systematic technical training…in Germany*. Kahlbaum was at the time head miller for the Adelaide Milling Company. Not only was he a miller but he was experimenting with wheat varieties in the search for those qualities which interested Farrer; in addition he had contacts with farmers who could try Farrer's wheats

In what seems to be the first letter dated October 6, 1896, Farrer was able feel he was in sight of achieving a result he had been seeking. It is evident that his interest was not now the overriding aim of his work.

> *The aim I have had before me for many years of making wheats for general cultivation yielding a stronger flour than such flour as the Purple Straws etc, is now likely to be successful as far as this Colony, at any rate is concerned*

The search for the best flour and thus the best loaf of bread was to be central to his work. F.B.Guthrie continued to provide the scientific authority to complement Farrer's work, and although Guthrie's mill was in operation it was still hand-worked and too slow for Farrer's impatience. Guthrie could not give as much time to the work because he was for a twelve-month period, acting

Professor of Chemistry at the University of Sydney, with a reasonably heavy load of lectures. Kahlbaum was to the wheat-breeder an opportunity for another source of advice and discussion as a practicing miller. Farrer judged that the times were coming round in his favour.

In 1895 with a bad harvest because of the drought, the Sydney millers were forced to import the hard Fife wheats from Minnesota. The added value of the subsequent flour showed the millers that bakers could make more bread from a sack of this flour than they had done hitherto with the soft Australian wheats. Now they were prepared to buy any local hard wheat of the type which Farrer was breeding.

In his letter of March 1897 he suggested to Kahlbaum that he come to Sydney and learn what Guthrie was doing. He gave the miller the advice he was later to give to Sutton, *the knowledge you will again will give you increased interest and pleasure in your work and in that respect alone will repay you for the trouble and expense.* Judging by subsequent letters he appears to have received a less than enthusiastic welcome from Guthrie.

The letters were exchanged on a monthly basis, right up until Farrer's death with, on Farrer's side, accounts of his progress and opinions on the nature of wheat varieties, as well as sometimes grumbles about Guthrie and Cobb. The April letter of 1897 showed the detail into which Farrer was prepared to go in his search for good flour: it recounted his testing of the grain for flavour, not by himself as he said his teeth were not up to it, but by his young assistant. *One or two the young fellow who does the biting for me, are pronounced by him to have a quite a pronounced flavour like that of pumpkin seeds. These I am inclined to think desirable, as well as some others which he describes as having a taste like that of olive oil, but still nutty and agreeable.* Based on flavour he was prepared to discard those wheats which did not measure up. *I do not care to distribute sorts with inferior flavour.*

In this man Farrer had a correspondent who was willing to pursue sympathetically milling trials for him. Kahlbaum became the confidant that Farrer needed. Farrer sent him seeds of unfixed varieties to try, telling him: *all the wheats sent to you will need fixing for one or two generations for your climate before a stock of them is made.*

By 1899 the letters were on a less formal level, ending *Yours sincerely* instead of the earlier *Yours faithfully* and chat about mutual acquaintances, while still discussing the qualities of wheat, such as White Essex and South Australia's White Tuscan which Farrer thought to be the same wheat and another, Sullivan's Early Prolific. Most of these discussions concerned the colour of the flour. Always careful to be able discard unacceptable wheat, Farrer wrote,

> *I am very glad to see you are going to have a testing mill in your Colony. It will be very necessary to keep it well protected from the dust; for I have found that the want of this precaution has done some damage in the past: at any rate, two cases have come to my notice in which the flour-colour has been spoiled from this cause: and at any time the detection of a grey colour in the flour is a ticklish job*

His relationship with Guthrie was not always free from tension. In 1900 when the pressure of work was heavy both for Farrer and Guthrie there appears to have been a disagreement over the matter of employing another assistant for the mill. Farrer as was his wont wanted to follow his path of ongoing regardless of restraint and have a man like Kahlbaum, *who has the knowledge of the difficulties the miller has to encounter which arises from differences in the milling qualities of the wheats which he is given to grind.*

At this time there would have been lack of departmental money but quite likely too, Guthrie might not have wanted such a man. The fact that Kahlbaum was German appears from Farrer's reference to the miller as having been trained in Germany, may have grated on Guthrie. On the other hand Guthrie may have been jealous of Farrer's sponsorship of Kahlbaum. Like many other things about these men it is a matter of speculation. The situation may have had a completely different explanation.

Farrer recognised himself that year that he was being regarded as a nuisance in the department, but nevertheless he wrote to Kahlbaum:

> *...I am sorry to hear that Mr. Guthrie has been giving expression to views which are so antagonistic to my own. It is true that if you want to put a boy who is not a perfect fool behind a mill, after a day or two's instruction he can grind a wheat and make it into flour; but that is quite a different thing from being able to give a reliable and valuable report on the milling qualities and milling peculiarities of that wheat... We want a man to have sufficient ability and technical knowledge as to be able to tell practical millers...what is the best manner to deal with that variety.*
>
> *A boy grinding wheats in a mechanical sort of way is doing no good for the country than could reasonably be expected for the wages he gets, if as much.*

An examination of the correspondence provides not only the number of wheats with which Farrer was experimenting, but his continuing pursuit of the ideal flour and the wheat which could produce it. The wheats which he was asking to be tested in the main were those older varieties which he was thinking of using for his own breeding purposes. Sending the wheat varieties to Kahlbaum and receiving some in exchange provided an avenue whereby it was possible to identify the varieties which were the same but existed under different names in South Australia.

Farrer used Kahlbaum to distribute his wheats for testing to interested farmers. Kahlbaum appears to have grown Farrer's varieties but whether on his own farm or like many others grew the varieties in his garden.

While his letters to Mark Carleton give an insight into Farrer's methods, enthusiasms and attitudes, those to Kahlbaum deal with the milling of wheat for the ease of the milling and the colour of the flour, with occasional expressions of annoyance if matters did not move as fast as he thought they should.

In 1905–1906 Guthrie again became Acting Professor while Liversidge was away for twelve months. Farrer was not one to make allowances. He wrote to Kahlbaum: *Your letter is the most welcome and the most important I have received for many a day. I have shown it to several others who are interest in wheat matters and all agree with me…It is a pity the baker should have given you so much trouble…I showed your letter to Mr Guthrie. I cannot understand why he should appear to take so small interest in the matter.* It appears that the bakers were using the wrong qualities of water to make the bread and were using litres instead of quarts with the strong flour wheats Kahlbaum had supplied. What Kahlbaum had established was that strong-flour wheats were liable to vary in the strength of the flour more than with the weak-flour.

A month later in Sydney there was a controversy aired in the *Daily Telegraph* over the use of strong flour and the amount of water needed to make the bread. Some people reckoned that a fraud was being committed on the customer.

ADVANCING SCIENCE

OVER THE WHOLE OF his professional life in Australia, Guthrie was an advocate of improved education for those involved in any occupation which could be said to be science related.

In 1898 he wrote on agricultural education in the *Agricultural Gazette*. He, like Farrer and Cobb, had decided views on how a farmer should be prepared by way of education for undertaking the practical nature of agriculture. He had a high ideal for education for farmers and considered that farming was becoming more of a profession than a practice based on ignorance and hearsay. He pressed for a system which would provide the would-be farmer, when he was a youth, with a training which was better than that which was regarded as good enough for the ordinary professional man.

He knew that no education could of itself be perfect in one sense, because it can never be complete, but it can be perfect in another, in method and direction, where nothing short of perfect would be admissible.

He posed the question which related to a general education: *What sort of education is it that turns farmers and mechanics into clerks?* To Guthrie the defects in the system were of less importance to a professional or commercial man than to a would-be farmer, because the non-farmer can remedy their faults at a university or bank. He emphasised the fact that often lectures full of dry facts had no real application in the real work of farmers.

He considered that to try and teach scientific subjects and practical farming in the same place would not work unless it was done at a great deal of expense with a large staff of teachers. The only way it would work would be if students had some prior training in scientific method. He proposed that there should be farm schools where, in addition to scientific subjects there would be classes in such skills as carpentry and farriery. A surprising inclusion in his proposals was the teaching of drawing–even before teaching writing. This needed the application of observation and the conjunction of mind and hand. He wrote that it was generally becoming recognised that practical subjects properly taught and especially scientific subjects had educational value equal to the study of language–as able to strengthen the mental faculties and having the additional advantage of being of value in later years

On the point of a number of subjects to be taught, *There need be, however, no cause for alarm on this score. As it is children are taught too few subjects. A variety of subjects gives relief to a child's mind.*

He stressed his main contention that scientific method which developed habits in observation and careful reasoning was better than the *massing of scien-*

tific facts and theory. To him all this study had a high moral quality where the student was taught to observe accurately; to reason carefully and dispassionately; to be honest and truthful; to judge impartially and to avoid making hasty generalisations. *He learns his own place in the scheme of things and acquires the virtues of self reliance and self-sacrifice.*

All this was of course directed at boys, but by that time there were girls schools which were following the same philosophy of education.

He followed his line of thought–an education which includes an elementary knowledge of science was desirable for both boys and men who wanted to consider themselves educated. The boy becoming a farmer, taught the scientific method would have the ability to understand the underlying principles of agricultural practices on which his success rests. However, he warned that these elementary principles of scientific method would not turn the boy into an expert entomologist, vegetable pathologist, or botanist. Guthrie was speaking in the light of his own experience of farmers who thought that they knew about all these fields.

He was proposing a course of education at the post primary level which would have seemed to many as being unnecessarily beyond the curriculum ahich the majority of parents then would have ever experienced.

In his years with the Department of Agriculture he was active in those institutions and societies which concerned themselves with service to society and his ongoing endeavour was to promote the best qualifications for the public good.

Later he was to carry the value of education and the public good even further when he became the president of the Royal Society of New South Wales and gave his presidential address in 1903. At the time he was acting professor of chemistry at the University of Sydney for twelve months. In his presidential paper he addressed the use, value and the future of chemists in Australia at the time. He noted that for the purpose of statistics in the census for 1901 they were divided into four categories: manufacturing chemists; analytical chemists and analysts; assayers and metallurgists.

Mining was then still the most important industry and assayers accounted for the greatest number,171, as opposed to manufacturing chemists, 96. Guthrie commented that chemistry hardly existed as an occupation in Australia considering the importance of science in manufacturing and industry. His point was that with the strengthening of existing industries and the emergence of new ones the value of the role of the chemist will increase.

The problem existed that the number of fully qualified practitioners was low and outside the laboratories which were in operation and where an 'apprentice' or pupil system was maintained, only the University and Technical Colleges provided any education in science in general or chemistry in particular. Because of this situation unqualified people could begin to practice and it would not change until there was a facility to set the standards of qualifications and the tests to show that the qualifications had been met. He cited the Institute of Chemistry in Britain as one such. The Institute had as its aim the promoti0n of better education of people wanting to qualify themselves to be public and private analysis and chemical advisers on scientific subjects; it set examinations and granted certificates of attainment; it sought to *elevate the profession of consulting and analytical chemistry by setting up a high standard of scientific and practical proficiency and by requiring on the part of its Members the observance of strict rules in regard to professional conduct.* The members of the Institute were in three categories: Students, Associates, and Fellows. An Associate needed to be over twenty-one and to have passed the preliminary examination in subjects of general education and also to have completed three years of theoretical and analytical chemistry, Physics and elementary mathematics. There were two further examinations to take in subjects as prescribed by the Council. To become a Fellow, the Associate was required to be continuously engaged for three years in the study and practical work of Applied Chemistry to the satisfaction of the Council. Guthrie, as a fellow, continued for many years to contribute articles to their Journal, having, no doubt, advanced in his attainments and reputation.

To strengthen his argument for following the Institute course, he drew attention to the fact that the Institute held examinations in local centres for those who could not travel. This could be done for New South Wales if the Public Service Board would accept the Institute's qualification as an essential prerequisite for an appointment as chemist in any government laboratory.

> *…it will be far preferable to take advantage of the machinery of an existing institution whose qualification is acknowledged all over the world, than create a new qualifying board whose decision will be discounted by reason of local jealousies and prejudices.*

The whole matter of qualifications and societies for chemists moved into the political arena with the registration of the Australasian Association of Chemist and Metallurgists under the Commonwealth Conciliation and Arbitration Act in 1904. The Australasian Chemical Institute regarded the AAC&M as being more in the nature of a trade union rather than a professional association.

Guthrie belonged to both and in his argument for a central scientific institute and a Science Department in government he was advocating the cause of science in general, both on the technical side and in the need for research to the benefit of the country. In 1916 the Commonwealth set up the Advisory Council of Science and Industry of which Guthrie was a member, with the aim of bringing into a national position scientific and industrial research. This Advisory Council was to become, in 1920 the Commonwealth Institute of Science and Industry and Guthrie was again involved.

Guthrie, in an article in 1918 on taking a stand for the establishment of the State Departments containing chemical research scientists, emphasised the value which could accrue to manufacturers if the other states followed the lead of those which had established laboratories for agriculture and mining departments. Although the acceptance of scientific research and application had made some advances in twenty years, he gave a wry summary of how the scientists of Agriculture Department were received in the 1890s:

> *There was just the same shaking of wise heads as one anticipates now. Certainly the farmers, who were to benefit the most from the future operations of the young department were among the most scornful of critics. The idea that a scientific man could be of any assistance to the practical farmer was ridiculous. Surely the one old cow knew better what kind of food was good for her than did the chemist with his balanced rations, and so the old cow went on 'blowing' herself on clover or ate immature sorghum and poisoned. herself on prussic acid.*

PRIDHAM

IN 1900 JAMES PRIDHAM was appointed by the Public Service Board to go to Lambrigg as Farrer's assistant. He was Dux of Hawkesbury College that year, and no doubt he would have been suggested to Farrer by George Sutton, then a lecturer at the College and himself working with Farrer on experiments with wheat at the College.

Pridham said in answer to the questionnaire regarding his knowledge of Farrer that the laboratory at Lambrigg was built that year just to the west of the wheat paddock. In addition to the laboratory, where the bunt[20] experiments were prepared, and Farrer's office, there was a room for the accommodation of the assistant. Pridham would have been the first to occupy it.

In an article which Pridham later wrote, he said that the most important quality Farrer wished for in an assistant was integrity. *If a man could no be relied upon for accuracy and honesty he had little time for him.* What Pridham most probably did not know was that Farrer had been put into an invidious position a short time before over his former assistant Franzen.

Farrer's working practices and discipline provided the base on which Pridham conducted his own breeding work. He recalled that, *the wheat plots were accurately laid out each season in beds, with paths at right angles…the wheat grains were dropped singly four or five inches apart in short rows spaced sixteen inches apart. Sowing was a cold job on a frosty morning…* Farrer's rule, *not to begrudge the time to pick up the spilt seed,* was more hon-oured in the breach than in the observance on those cold Canberra mornings.

It was not just the planting and caring for these hundreds of single plants which would engage the new recruit, but it was the recording. Like thorough-bred horses, the pedigrees, were, he said, *in many cases of such length that small handwriting was essential to enter them in the field notebooks.* It was these meticulous notebooks which were so remarkable in Farrer's work, even though there no other remarks or accounts of how he proceeded. Pridham absorbed Farrer's dic-tum concerning the prime importance of facts and observation, taking all he had learned into his own career as a wheat -experimentalist, and later plant breeder.

Of all Farrer's 'apprentices', with exception of George Norris, Pridham was the one who was able to carry Farrer's ideas forward. He went on to produce both new varieties of wheat but also new varieties of oats.

After leaving Lambrigg, Pridham went off to the Bathurst Experiment Farm, where he began oat breeding, convinced that, like wheat, improvements could be made in the quality of the then-existing varieties so that they could be grown, like the new wheats, in the hotter, drier parts of the country He was to go on breeding wheat and oats to produce varieties which became included in the seedsmen's catalogues.

Following a time at Bathurst, he went as wheat experimentalist, to Longeranong Agricultural College, not long after it reopened in 1905. Situated in the Wimmera, with its wheat experiments related to varieties for drier areas, he carried in his experimental baggage ideas gained from Farrer and his experiments at Coolabah. As an associate of Farrer whose Federation wheat was to save the Wimmera, he had the right credentials for acceptance in that part of Victoria.

He returned to New South Wales in 1911 to a position at Cowra Experiment Farm as assistant manager to George Sutton, who not long after left for Western Australia, leaving behind him an unfixed variation of Federation. Sutton had noticed among Federation in the wheat breeding plots that there was a plant which did not conform to the usual description. It had hard translucent grain where the rest had the soft opaque grain of Federation. This variation would have delighted Farrer, self-sceptical as he had been about the grain of this wheat. Some discussion must have ensued about the likely parents of this plant. There is no indication of what Sutton may have thought but Pridham was of the opinion that it was a cross between Federation and Comeback. He set about fixing it, later running trials at the Wagga Experiment Farm. In 1914 he honoured it with the name Hard Federation. While for a time it proved popular, it later became susceptible to both stem and leaf rust. It was introduced into the United States where not only was it grown commercially in California but became a popular on a number of experiment stations by 1937.

After Sutton went to Western Australia, Pridham was reclassified as plant-breeder instead of wheat-experimentalist, acknowledging his wider role. While he continued with wheat he also began to make a name for himself as an oat-breeder. In both strands of breeding he relied on Robert Hurst at the Wagga Experiment Farm, as had Cobb, Farrer and Sutton. By Pridham's time Hurst had been watching and working with these men who had been among those laying the groundwork for a wheat industry about to take its place in the world. Two of his wheats Wagga 14 and Wagga 21 were introduced in 1900 and 1901 respectively so he must have been working on them in Cobb's time.

One of Pridham's wheats, Bobin, has an added interest besides its parentage, a cross between Thew, (one of Farrer's earliest wheats of 1890), and Steinwedel. Mrs Farrer is recorded in his annual report of 1911 as having managed the cultivation of Bobin at Lambrigg. It went on to become the third leading wheat in NSW by 1934.

In 1916 Pridham wrote *Wheat-breeding in New South Wales* put out by the Department of Agriculture as a Farmers' Bulletin. In the introduction he said in different words the sentiments that had been written and argued before:

The average farmer pays considerable attention to the breeding of his livestock, and will have nothing to do with a mongrel or a weakling, but gives scant atten-

*tion to the character of seed he sows, provided that it is not smutty or unduly
shrivelled. We may secure good results by suitable preparation of the soil and
manuring, but it is impossible to secure the best results from inferior seed*

He went on to explain the difference between breeding wheat and breeding animals. *It is easier and quicker to propagate plants than it is animals and wheat
lends itself moire easily to improvement than do close or cross fertilised plants where two
individuals are concerned in seed production. Wheat, however, although nominally self-fer-
tilised is not quite the uniform product that one would expect. Every grain of Purple Straw
wheat does not grow a plant exactly similar to its neighbour in the same way that Jonathan
apple trees bear similar fruit*

Pridham appears to have continued to breed wheat until 1927. Continuing
with his oat breeding, he produced a number of new varieties such as Mulga,
Belah and Buddah. His work was not only important for the southern region of
New South Wales but also for other States and in the United States with his variety Bond

Robert (Bob) Hurst had joined the Farm as a workman in 1892 in John
Coleman's time. At first he had helped Nathan Cobb but later Maurice McKeown
reluctantly assigned him to William Farrer at those times when Farrer visited the
Experiment Farm. To McKeown's annoyance Farrer also wanted him to follow
through the work when the experimentalist was not there. William Farrer was
always prepared to acknowledge Hurst's skills and did so in a letter to George
Sutton, admitting to some mistakes about which Bob Hurst had corrected him.
Farrer recognised his worth, when in his 1903 report he wrote: *The greater part of
the work was done by Mr Hurst to whom I furnished a list of crosses which I consider likely
to have some good results. Mr Hurst is now an expert in this work.* This was evidently not
something which Maurice McKeown was prepared to acknowledge.

When Farrer died in 1906, Farrer's collaborator George Sutton took over as
the Department's chief wheat-breeder working out of Cowra Experiment Farm
and he continued Farrer's practise of leaving the working of the plots at Wagga
to Bob Hurst. In his reports Sutton acknowledged Bob Hurst's expertise and
referred to him as Mr Hurst, and with this honorific giving him a dignity not
accorded to workmen at the time. On the other hand McKeown was seeking to
discharge Hurst. The following letter written to Sutton in August 1907 was that
of a man losing patience:

*I hope you are keeping in view the arrangements we made re finding a successor
for Hurst, as I had occasion yesterday to 'shake him up' for indulging in his old
habit of 'yarning' at the fence…the workmen and thus wasting the time of both
persons concerned.*

*I have instructed Mr Carne, our experimentalist, to report to me in the
future should I again have occasion to reprove him. His dismissal will follow
without future notice.*

Pridham followed Sutton as Departmental wheat breeder, also supervising the work at the Wagga Experiment Farm, leaving Bob Hurst to continue to work on wheat and barley. This man, by long experience and knowledge of the methods he had acquired from Cobb, Sutton and Farrer had become an expert.

In his youth Bob Hurst had had experience of farming as his father John Hurst had been one of the earlier farmers in the district. Although he was an old hand on the Experiment Farm, for some reason, perhaps because of his association with Farrer, Maurice McKeown did not actively support him in improving his status in the Department. In 1907, he wrote to the Department:

referring to the position of the Experimentalist which under present circumstances it is difficult to fill satisfactorily, I have the honour to ask that the matter be left in a similar position to that of the general farm hand. These are engaged on a daily rate by the manager and I am satisfied that more satisfactory work can be obtained under this system than under that hitherto carried out.

It can only be supposed that with the new structure, McKeown had taken advantage of being able to keep a tighter control on non-professional staff, which from Valder's day had always been employed on a temporary basis[20]. This temporary basis of Bob Hurst's employment was to stay until he retired, forgotten in the departmental depths. The result was that in spite of petitions made by him and on his behalf, he left the Department of Agriculture in the late thirties with nothing but a watch!

Bob Hurst was responsible in carrying out Farrer's experiment instructions. In the letter to the Under-Secretary where McKeown put his side of the case which Farrer had brought against him, he had this to say about the work Hurst was required to do:

Hurst has to supervise and assist in the labour of sowing and harvesting, keep voluminous notes, mark the plants by a tedious process on tying on the various coloured tags, keep the plots clean.

He developed many wheat varieties of his own. In the time Bob Hurst was assigned to William Farrer at the Farm he always kept a watchful eye on possibilities among the plots.

In 1901 he had fixed a cross between *Federation* and *Volga*[21] which in 1911 was named *Canberra*. Working on *Canberra* he developed *Duri* at the Farm, naming it in 1922. Carrying on under Pridham's supervision Hurst evolved the variety *Waratah*, which proved to be very popular with the farmers. In 1925 it was the sixth most popular variety and by 1929 was the leader. These were not the only varieties accorded to Hurst, but the most successful.

Of all the identities from the early days of the Wagga Experiment Farm, Robert Hurst was the one who made an impact as far as the students of the 20s and 30s are concerned, and from the point of view of the students, there was one important person who was able to train them in the precise nature of experimentation and the care which was needed was Robert Hurst. By this time he made his own contribution to the development of new wheat varieties

Ted Dixon had had close contact with him when he was a student at the Farm in 1923–4. In those days of being allocated to different sections of the Farm for training, students would often swap with somebody if the section to which they were allocated was not tho their liking. Dixon was able to do this so that he could work with Bob Hurst and his wheat breeding.

In spite of his successes Bob Hurst was not to reap any rewards. In a letter he wrote to *The Daily Advertiser* in June 1935 reveals his hurt and bitterness:

Sir,

I was greatly interested in your leading article in last Saturday's issue under the heading 'Rewards for Wheat breeders'. As I was associated with the breeding of wheat, oats and barley at the Wagga Experiment Farm for nearly 40 years I would like to make a few remarks on the subject.

In 1894 I started this work under Dr. N. A. Cobb who was at that time experimentalist for the Department of Agriculture. I continued with him until he resigned from the position and went back to America. Mr Farrer was then appointed to take Dr Cobb's place. From that time until Mr Farrer's death, I did practically all his work in connection with the crossing of wheats etc. and the selection of plants carried out at the Wagga Experiment Farm. As Federation wheat was mentioned in your article I would like to state that the year this wheat was named (1902 I think it was) we had three strains of this crossbred growing at the Farm. The strain that I considered best was chosen by Mr Farrer and named Federation. After Mr Farrer's death Mr George L. Sutton was appointed wheat experimentalist.

During the years I was at the Farm I made many crosses in wheat oats and barley and raised some very good varieties. I also made a lot of crosses listed from other Government farms. When the cross seeds were harvested they would be forwarded to Cowra Farm for planting there. Among the new wheats raised by me were Canberra, Waratah, Riverina and many others. Nabawa was also raised from a cross made at the Wagga Farm. When Mr Sutton (who was then in Western Australia) received the seeds of these crossbred it had been growing for three years at the Wagga Farm. When Mr Sutton finally fixed this type, he gave it the Western Australian name Nabawa.

Algerian was also a variety raised by me. It is a cross between Old Algerian and another variety called Red Rust Proof. When it was finally fixed and proved to be a good oats, Mr Pridham (who was over me) suggested that the new oats be called Algerian and that the Old Algerian be rejected. I did not like that and stated in a letter to the manager of the Wagga Farm that 'if the department had bred an oat that was better than Old Algerian, it was worthy of a name and let the farmers decide if it was better than Old Algerian or not.' However, the new oat was called Algerian, and Old Algerian dropped out, no one getting any kudos for the new variety.

During the whole of the time I was at the Wagga Farm I was kept on the temporary staff and therefore not allowed to pay into superannuation, which would have enabled me to provide for my old age. I was retired from the Farm nearly three years ago, on reaching the retiring age and all that I got was long-service leave. I have approached the Minister for Agriculture Mr Main) through our member (Mr Kilpatrick) to try to get come recompense for my long service, but I was told they had no money. The last communication I had with Mr Kilpatrick on the matter, he told me that I had no hope of getting anything. My savings out of my wages were practically nil as I had seven children to rear and educate. For about twenty years I was paying off on a place for my own home. This property is now mine and valued by the Wagga Municipal Council at more than £1000. But what good is it to me? Eventually I will have to sell and when most of the proceeds are eaten up I will probably be able to get a few shillings a week from the old age pension fund to enable my wife and I to exist. This appears to be my reward for nearly 40 years service in breeding new wheats and raising pure stud seed at the Wagga Experiment Farm.

He had continued at the Farm until he retired in 1932 using his expertise in the identification of wheat for the benefit of students and any others who were interested.

Bob Hurst made a lasting impression on John Waugh, a student 1929–1932, which he was able to recall sixty years on,

Bob Hurst: an expert on wheat-breeding and assistant in the plant-breeding wheat research laboratory. As students we would take a few grains of wheat from say 15 or 20 bags of wheat from local bred wheat or imported wheats from Egypt, Canada, Russia, Argentina etc. and while Bob Hurst was having lunch we would put 50 or 60 grains of wheat (after listing which bags they came from) on Bob's desk and ask him to identify them. He would pick up his ruler and draft the grain out in respective heaps and tell us which bag they cam from. He was truly an expert in his field and worked with Mr Farrer in the early days producing Federation and various other types of wheats to the benefit of the Australian wheat growers.

Sadly, in the end, all that this long-standing servant of the Crown had for his years of service was a watch. It was not until the opening of the Wagga Agricultural College that there was an acknowledgment of his important place at the Farm and Australian agriculture, when he was accorded the honour of being asked by the Minister for Agriculture to plant a tree at the opening ceremony.

ROWLAND BIFFEN

ROWLAND BIFFEN WAS one of the bright young Cambridge graduates who became enthusiastic over Mendel's theories in 1900 and saw them as the template which not only accounted for variation but could be used it to make their own genetic modifications.

Biffen was born in 1874 and entered Cambridge University in 1893 where he studied natural science. By 1898, after a trip to the Amazon to explore for sources of rubber and in appreciation of the industrial potential of botanical species, he began to specialise in economic botany. In 1899 he became a lecturer in botany in the new School of Agriculture at Cambridge. In his position as botany lecturer he realised the importance of Mendelian theory in relation to the role of agriculture in the British economy and the State's role in having governance over the food supply for the country's increasing industrial population, especially in the light of any threat from a foreign power. Britain's agricultural landscape had changed with industrial growth and new conglomerations of people.

A letter written to Farrer by Biffen on January 9th, 1905 brought a new enthusiast into Farrer's circle of correspondents and there cannot be any doubt how it would have heartened the old man.

Dear Mr Farrer,

I am at last posting you a copy of the paper I promised you—the reprints have only just reached me so I trust you will pardon the delay.

Since writing to you I have had along talk with Humphries with regard to the water absorption by flour and 'strength' The gist of it is this. A strong flour, to us, is one which will produce a well-piled loaf- ie. a light, high one with a good surface and no signs of squatness and heaviness. From the amount of water absorbed on making the dough we cannot say whether a flour is strong or not. In this sense of the word, for the baker to get the best results from his material has to make dough of different 'tightness'. If for instance a baker made up Hungarian flour into as tight a dough as his experience has taught him is right for English inland flour he would not get satisfactory loaves of the maximum size or the maximum amount of bread.

Further Hungarian flour will make a very large number of loaves per sack, probably more than any other flour we know of, but, though strong, it is surpassed by many other varieties in strength. It seems then that the yield per sack and the capacity to make big loaves are not necessarily combined in the same wheat. The Indian wheats and many of the Russian durums give the same result—they will absorb enormous quantities of water and yield many loaves to the sack but they never give what the English baker would call a well-piled loaf.

If you think this conception of strength is not the right one please overhaul it for us for we are anxious to make the best progress we can in the matter.

I have read through the paper you were kind enough to send me and I think you will find-that allowing for the different standpoint now possible in plant breeding—our experiences agree remarkably well. Your 'clumps' interest me greatly for I have been investigating the same phenomenon for some time. So far my statistics lead nowhere—as far as I can see, but it is in these clumps particularly which give rise to the spelt-like ears. In spite of this, and in spite of interest their spelt like foliage I am not convinced that they are to be regarded as reversionary forms. I know of similar phenomenon in other cases and hope to be able to worry out a reason for them which will stand testing. It's so difficult to be sure that the primitive wheats are spelts.

With kind regards,
Yours very truly,
Biffen

Another letter was written on January 9th 1905

Dear Mr Farrer,

Very many thanks for your letter and the copy of your paper which I have received today. I knew of your work with wheats but I have never had the opportunity of realising how comprehensive it was and I find that I have done you a considerable injustice in a paper now in the Press by not referring to it in great detail. I will make up for that as soon as I have the opportunity though, and I will forward the offending paper to you in the course of a week or so. You will see from it that we have for some time fairly definite data to work with and our old bugbear of the 'fixing' of varieties originating by cross-breeding has disappeared, thanks to Mendel's work on heredity.

Your work on breeding plants immune to Tilletia interests me in no ordinary fashion for I have been gathering data together for several years as a preliminary to attacking the problem of raising varieties, not only of wheat, but of our agricultural plants in general, which should be immune to their respective parasites. I think I have got fair evidence, not as good as I should like it to be, that the liability to Glumarum and immunity are differentiating characters in the Mendelian sense. If so, then our problem is immensely simplified. I shall await the results of your experiments with eagerness, especially if you have managed to do this by cross-breeding.

With regard to the milling side of the problem I am very fortunately situated. Humphries, to whose words you refer has an experimental milling plant, a trained baker and a most extraordinary 'eye' for quality man ever had—all these I draw on fully. I think I am justified in saying I have succeeded in raising wheats of good quality and cropping powers already, though we still have to raise them in bulk in our work the only reliable index of 'strength' so far as the volume of the loaf the flour will make and we now take as our definition of 'strength, " the capacity to yield well plied loaves". The water-holding capacity seems less importance to us than the gas-holding power of the gluten so our final test is in the bakehouse.

Generally speaking I should say that the strong wheats do take up more water than the weak ones when worked in a standard dough, but the matter is complicated to a certain extent by the amount of moisture already present in the grain—with your dry climate this difference would I take it be practically negligible. One of our tests where Square-head Master, a distinctly poor wheat, was compared to a really strong Fife showed that the latter only made about one quarter loaf more to the sack—but no one would have cared to have eaten the Square heads loaves.

I shall be seeing Humphries shortly and will talk to him about this part of the work and write to you further on the subject. I will also send samples of any such wheats I can lay hands on.

We are getting together some extraordinary results with regard to acclimatisation—which at present are rather inexplicable. Some of the Fife strains imported from Canada are practically so strong now, after four years cultivation, as they were when imported, but on the other hand several of the best Hungarian wheats are now no better than average English. At the same time their yielding power, especially in poor soils seems to be steadily increasing. How long will it last I don't know. I am now working on the inheritance of the total nitrogen content of the grain. It is a wearisome business, but I hope in the end to get some information as to how strength is inherited on crossing weak and strong varieties together. The total nitrogen content seems to be a fairly reliable guide as long as one confines oneself to bread-making wheats.

Cambridge must have altered a great deal since you were up. Even within the last twelve years it has grown out of all knowledge. Claremont is near the railway station. I have moved out here so as to be near the experimental farm of our agricultural department. These hybrids want so much attention that I found the labour of getting to the farm too great and too time-robbing

with kind regards,
Yours sincerely
R.H.Biffen.

The correspondence continued with the following letter:

Dear Mr Farrer,

I have sent a few samples of wheats off to you today. You will find that Fife x Lammas and Fife x Rough Chaff is represented by several distinct types. They are, I think, good quality. I am also sending you a paper by Humphries which will explain itself. In it you will see he gives his reasons for defining 'strength' in the way I have already explained to you.

I've extended my cages for this season's work. Sparrows are awful in spite of poison and small shot.

We are growing on a long series for milling tests this season.

We are very busy planting now..

With Kind regards,

Yours sincerely,

R.H. Biffen

The following extract is part of a letter for which I do not have the first page:

After this harvest I will send you a number of really strong Russians which we have isolated from these graded wheats. Unfortunately what little seed of them I had is in the ground now. We are hoping that some of them will give us better quality than Fife. So far and this is their fourth season in this country they show no signs of deterioration and now we expect from analogy with other cases that they will hold out indefinitely. The Romanian wheats have all been rejected as unsuitable for our conditions and I have got any by me at present. I shall be sure to come across some though and I will secure them for you. As a rule they are fine looking varieties but they lose all their strength in our climate. Two seasons was sufficient to pull them down to the level of the ordinary English varieties.

I am inclined to think that on the whole water absorption is not at all a bad criterion of quality especially where one is dealing with one form of endosperm only, such as Fife. Our difficulty was that on tasting the endless number of types that came into our hands we found that it frequently gave us different results to the baking tests so after endless attempts in other directions as well we were driven to conclude that the oven was the sole thing we could rely on. We are still groping about to find some easier method of determining strength as it is obviously an Herculean task to bake, let alone grow on, all the hybrids which appear to be of promise. I now have a student at specialising on the proteids of wheats in the hopes of getting at some method for testing quality in small samples.

Those grass clumps are extraordinary things. I had no idea that they had been described until you sent me your paper. I find them in particular abundance where Rivett wheat is one of the parents so I have sown a considerable number of these this season to see if I can get at the bottom of the mystery. There is sure to be some Mendelian explanation of it. Your spelt wheat may be a mutant but it is very difficult to be sure of it. There always seem to me so many possibilities of stray seed creeping in even in the best managed of experiments that I should be as critical as possible of it. The most likely explanation which I can think of is that a natural cross occurred last season. In that case your plant would be an F1 of spelt parentage. The result of crossing ordinary wheats with spelts as far as I have had any experience—and this is not much— are that

*such plants are more or less spelt-like. They have closed spikelets and brittle
rachis for instance. If this is the case you will get some evidence for it in the next
generation. Should it breed true though and there is no likelihood of stray seed
occurring it will, be strong presumptive evidence of a mutational origin. The
quotation from your American friend's letter is very interesting though the case
he describes of a durum arising from 'emmer crossbred' wheat does not strike me
as conclusive. On the face of it I should fancy it is simply a re-arrangement of
the characters already existing in the parents not a de novo origin. The general
form of the ear would be provided by the bread wheat and the awns and keeled
glumes by the Emmer. The latter would also give a grain shape very similar to
that of the durums. If he had got a type like T. Polonicum from this cross it
would have been another matter, for the long glumes could not have come from
either parent. Looking at the question now I see no real chance of getting any
satisfactory data from a study of hybrids as to the origin of our cultivated
wheats. I have several curious specimens but by which? I could call reversions
but in the absence of any positive evidence one way or the other I do not dare
to…*

Biffen continued in another letter on the matter of inheritance:

*These have all occurred in the second and third generation and so far I have
never come across a case in the first generation which I could not explain satisfactorily on Mendelian grounds.*

*In the case of sweet peas the first generation between very diverse parents for
the old type known as Painted Lady or the old Purple is often a reversionary
form which crops up.*

*The following generations though are the ordinary Mendelian type of segre-
gation. It is all a big puzzle yet but we shall win through some day!*

*I hope the wheats I sent you have reached you by now. The mails seem to
have been playing all sorts of tricks lately. One letter from a correspondent in
Victoria took six weeks to reach me.*

*I see on looking this through I have made bad work of the spelling here and
there. Please excuse it for I'm writing with a machine in which the key board is
not very familiar.*

Kind regards,
Yours very sincerely,
R.H. Biffen

There is a written postscript.

*By the way I have just heard from Dr Saunders. He has sent a budget of reprints
which I must sit down to some evening. I see that he has a hybrid between Polish
x Fife—from the figures very similar to my Polish x Rivett.*

The following letter was written on the 11th January 1906.

Dear Mr Farrer,

I am writing post haste to catch this mail. I will write in more detail later.

You will find a very readable account of sports and mutations in a new work by deVries called 'Species and Varieties, their origin and mutation'. The open Court Pub.Co. Chicago. 1905.

I hope the wheats have arrived by now.

Yours sincerely,

R.H.Biffen

The following letter was written in 2nd February, 1906

Dear Mr Farrer,

Your long letter came to hand a week or two ago so I am answering it before the rush of spring sowing begins. What with the hybrid barleys as well as the spring wheats we shall soon be busy now.

I had already found that the Zealand wheat had a higher nitrogen content than your figures indicated. We had estimated the total N. to see whether it would increase under our conditions. Most wheats as you know deteriorate a good deal under our conditions. In spite of the high figures the wheat is remarkably weak. If chewed it hardly leaves any residue in the mouth and that is fairly rough and ready test for strength. We discovered the peculiar action of sulphur dioxide on wheat which you mention quite accidentally. There was a plague scare here a year or so ago and a cargo of wheat on a suspected vessel was fumigated, with disastrous results as far as the value of the grain was concerned. On attempting to wash out the gluten we only got a gummy mass which ultimately worked its way through the washing silk. I forget now how it baked but I believe very badly.

By the way the term Gulka wheat is a little misleading. The name is given to a graded wheat which usually contains some seven or eight varieties. The one I sent is the pick of these as far as our conditions are concerned. If I can get hold of a trade sample as received in this country I will send it to you.

After this harvest I will send you a number of really strong Russians which we have isolated from these graded wheats.

Biffen by 1906 was regarded as the foremost wheat scientist in England and was consulted by Albert Howard who had been sent from England to India in 1905, in a botanical advisory capacity to the Indian Government. Cambridge graduates, Howard and his wife had been brought to India to take charge of the development of agriculture. They were botanists by training but knew nothing about wheat.

The British government had made wheat a priority following Crookes airing of the wheat problem which would arise in the event of war. Politically the colonies of which they considered India one, were to supply the raw materials and Britain was to provide the products of industrialisation. India at the time was only sending 10% of her wheat harvest to England.

The Howards found themselves having to develop a wheat-breeding pro-gramme because they decided that what existed in India in the way of wheat breeding was too haphazard and unproductive. They established a wheat insti-tute at Pusa.

In a book later by Louise Howard on the work of the Howards, she wrote: *Wheat was so important in India that it would have been strange if there had been no pre-vious investigations. These however had been misdirected and for this reason had been unsuccessful*

> *...An obvious duty was the examination of the existing types of Indian wheat and improvement by plant-breeding. Previous investigations on this point had been desultory and local and had committed the unpardonable error of trying to solve the sinuation by short cut, namely, the introduction of foreign varieties. These had proved a complete failure.*

Farrer since the days of Bancroft exchanged wheats with stations in India and W. H. Moreland had, in 1900, been sent to Australia to study Farrer's meth-ods. Now ironically, the Howards sent some Indian varieties to England to Biffen to be crossed and trials made. Farrer's dictum that climate will affect the way a variety behaves proved true with Biffen crosses. In India they were unproductive.

In 1916 the Imperial Economic Botanist, stationed at Pusa, sent a parcel of Pusa 4 to New South Wales Department of Agriculture. Dr Martin-Leake who in 1906 who was in charge of Cawnpore Experiment Farm as W. H. Moreland had been, wrote as a note to Louse Howard's book that he and Howard had inspect-ed the plots which had been planted of the wheat which Moreland had received from Farrer, which he said had been crudely crossed before Mendelian princi-ples had been applied. He said the basis of the crosses was putting successive 'doses' of Australian wheat on Indian wheat. In the event he and Howard made selections at both Pusa and Cawnpore and selected from the crop. The outcome was a series of varieties called Pusa.

> *If my interpretation of what happened is correct and I can see no other ...Australia was merely importing one of the wheats originally imported into India by Mr Moreland. It is believed that Pusa 4 is related to Federation, a gift from Farrer to Moreland on his visit in 1900.*

In examining the sources of Farrer's various varieties they reveal their Indian origins from Etawah on. The wheel had made a couple of turns: Indian wheats into Australia and the new varieties back to them and Howard's consult-ing Biffen who was well aware of Farrer's methods and ideas.

Farrer wrote to Biffen in March 1905:*in your letter you speak of 'the old bugbear of fixing varieties' This work for the last twelve or fifteen years has given me no trouble whatever. It seems to me from what I can see of Mendel's theory of heredity, that the consideration I then gave to the matter of fixing varieties led me to adopt the system, which, for all practice purposes, Mendel's theory indicates as being the best…The practice was adopted from what appeared to me to be commonsense considerations; I certainly did not have Mendel's theory to work upon.*

On the 14th April, 1905, he wrote: *There is one point in connection with Mendel's law that it seems to me not to provide for. It is when varieties, which differ sufficiently in type, are crossed, the variable generations seem to produce individuals which differ in all the qualities in which varieties differ: e. g., by crossing two late sorts of different types, it is quite possible to get early sorts. I cannot recall just now an instance in which I have got a very early variety in this manner, but I have made a such crosses varieties which are distinctly earlier than either parent. Mendel's Law, I fear, is not likely to be of great use to me in enabling me to improve my methods, because in nearly all the crosses I make, one of the parents is an unfixed crossbred, and frequently a plant of the first generation from the cross.*

At the Winnepeg, Canada, meeting of the British Association for the Advancement of Science Biffen referred to Farrer's work *as the only successful wheat-breeding experiments yet undertaken in the world.*

VALE COBB

The number of contributions made to the *Agricultural Gazette* after Nathan Cobb's return in 1900 were far fewer than before he went overseas in 1898. He concentrated on the instalments of his work on nomenclature of wheat published in the *Agricultural Gazette* from 1901 and later gathered into one volume, *The Universal Nomenclature of Wheat,* complete with five pages containing 1000 entries in the index.. Campbell had denied him a return to anything to do with wheat but according to Richard Sayre, in *Art in Phytopathology,* the publication of his opus made him recognised as a world authority on wheat. In the meantime, he published the results of his work on the sheep fluke while continuing in his initial interest, the study of nematodes, especially in tropical agriculture.

In 1903, the United States Secretary for Agriculture asked Cobb to establish a Department of Agriculture in the newly acquired territory of the Philippines which declined. Frieda Cobb Blanchard in her biography of her father said that the Secretary was anxious that the United States policies regarding agriculture in this newly acquired war prize should be established on the right lines. Cobb was flattered but refused being conscious of the need for his children to be educated in America. The following year he sent his son Victor to study at the Worcester Polytechnic Institute, from which Cobb himself felt that he had received a good education.

By this time Cobb, besides considering the education needs of his children, may have been becoming disenchanted with the pace at which the New South Wales Government was dealing with the new advances in matters such as irrigation. He took his four months long service leave and sailed for the United States to make personal contact with men associated with the Department of Agriculture. Erwin F. Smith one of Professor's Galloway's staff in the Department of Agriculture and a colleague of Mark Carleton's recommended Cobb's appointment to the Department. He had apparently been impressed with Cobb's work on gumming in sugar cane and no doubt his work on wheat. However, before his Government appointment he took up a two year appointment to establish a research station for the Sugar Planters Association in Hawaii, with the title, Director of the Division of Pathology and Physiology.

He and the family returned to Australia preparing to take up his new task in June or July of that year. When he resigned he took Grosse and Chambers with him. A week after his resignation was announced he gave an interview to *The Sydney Mail* at his home in Pymble. It had nothing to directly with his work, but it was about efficiency in travel. He was fascinated with the speed of the changeover time for passengers on the ferry to the train in San Francisco. He told the reporter how he had timed the operation at each stage. There was a sense of implied criticism about the lack of attention to the importance of efficiency and concern to provide people with the services they needed. The interview highlighted Cobb's almost obsessive interest in what would work in the most efficient and time-saving way. This interview was given in the same year that an extract from a letter was printed in the *Agricultural Gazette* berating the lack of commitment to the concept of bulk storage of wheat and the uselessness and inefficiency of bags for storing grain against mice and other vermin.

Cobb was feeling frustrated in whatever efforts he was trying to achieve excellence in research, organisation and agricultural education.

At the time he resigned the big news in the rural field was the matter of water conservation and irrigation and whether it should be privately activated or government run. Irrigation had become of interest since the Chaffey brothers had begun their implimentation at Mildura in 1887, and had been promoted in New South Wales over the years by Leopold de Salis in the 19th century through letters to the newspapers, having seen the possibilities along the Murrumbidgee from the time of his holding at Junee in the Riverina from the mid-century. Cobb in his interview voiced some implied criticism of the Government on its inertia in the matter:

> *I was particularly impressed with the prospects of the Pacific Coast. In the near future electric powere will be delivered 400 and 500 miles from the power house and ...power will be available in this way at a lower price than it could be supplied by steam, even if the coal could be put down in San Francisco at a price it now brings in the mines. It seems almost beyond doubt that the Rocky Mountains and the Pacific Coast will advance more in population and wealth in the next quarter of a century than any other part of the world. Electric plants will spring up. Immense irrigation schemes will be brought into existence, in fact are now in construction, one st least being ready for crops. That is the Truckee area. Already a Government fund of over 20000000 dollars exists devoted solely to this purpose. The sale of lands to be irrigated is looked to augment this fund enormously. American scientists are turning attention particularly to the solution of the problems of semi-arid regions, thus far with gratifying results. The magnitude of interests is calculated to enlist the services of some of the best minds in America.*

This was not the first time Cobb had taken the Government to task. In the early days of the department he had attacked the lack of proper accommodation for the officers on the farms, whom he said were forced to live in tents of in other sorts of makeshift shelter. The implications which lie in the report echo what Farrer in 1873 wrote in *Grass and Sheepfarming* while comparing what was happening in New South Wales with the agricultural progress in the United States. *Here a morbid, overstrained and generally misdirected policy of economy seems to be paralysing all true progress*

SOILS

ART OF GUTHRIE'S brief from the beginning of his employment with the
Department of Mines and Agriculture had been the analysis of soils. From his
meetings with Farrer he would have known of Farrer's pamphlet, *Grass and
Sheepfarming* and its discussion of the constituents of soil, from anecdotes
recounted by Farrer about his surveying days questions would have been asked
about the kind of country the West Bogan was, and because of Farrer's conviction that the western country could produce wheat, Guthrie's interest would no
doubt have increased over time as his tasks increased. From Farrer he would
have heard of E.W.Hilgard, one of the pioneers of soil science in the US, with
whom Farrer had been in contact in the days of exchange of wheats with Blount.

In 1899 the United States Department of Agriculture had begun mapping
a country-wide soil survey under Milton Whitney, with whom Hilgard had a series
of intellectual battles over the matter. Milton Whitney had conceived the idea of
mapping soil characteristics as a means of promoting agricultural development.
Early soil surveys had been made to help farmers locate soils responsive to different management practices which would be the most suitable in relation to the
kinds of soil on the farm. Hilgard's argument with Whitney was because the latter's disregard of a soil chemical analysis in the survey.

Hilgard's conclusions relating to the survey he had done in Mississippi was
contained in a report he prepared in 1860:

> *My exploration of the State had shown me such intimate connection between the
> natural vegetation and the varying chemical nature of the underlying strata
> that have contributed to soil formation, as to greatly encourage the belief that
> definite results could be eliminated from the discussion of a considerable num-
> ber of analyses, of soils carefully observed and classified with respect both to their
> origin, and a comparison of the data with the results of cultivation; and that
> thus it would become possible, to do what Liebeg originally expected could be
> done, viz. to predict measurably the behaviour of soils in cultivation from their
> chemical composition.*

Hilgard over the next 30 years promoted concepts which were the three points
for the basis of the plan of the soil survey: judging the land by its natural vegetation;
examination of the soils in the field to reveal the depth of the soil; chemical and
physical analysis of the soil collected in the field, stressing the need to understand
the plant-available elements. Whitney on the other hand emphasised soil texture
and the ability of the soil to provide moisture and nutrients for plants.

When Whitney began many of the workers in the field were geologists as they were skilled in the required field techniques and in the scientific correlation which were most appropriate in the study of soils. The accepted wisdom with the geologists was that the soil was, in the main, the result of weathering of geological formations, seeing soils as disintegrated rock In the laboratory, the balance sheet theory of plant nutrition was dominant. With them the soils were rarely examined below the depth of normal tillage, in contrast to Hilgard's views. The balance sheet theory of plant nutrition was accepted in the laboratory, while the geological concept guided the field work sll into the late 1920s.

Hilgard had gone to Berkeley in 1875 where he expanded the College of Agriculture and the experiment activities of the State of California. By 1884 he turned his attention to the idea of a national agricultural survey, having already completed regional agricultural maps such as the one for Mississippi.

The United States was not alone in making a survey. The Russians had begun in 1870, but the Russian workers saw the soils as *independent nature bodies with unique properties resulting from a unique combination of climate, living matter, present material relief and time.* What the Russians were developing as concepts required that all the properties of soils be taken as a whole as a completely integrated body.

Milton Whitney had conceived the idea of mapping soil characteristics as a means of promoting agricultural development. Early surveys had been made to help farmers locate soils responsive to different management practices which would be most suitable with relation to the soils on the farm.

With Guthrie and his farmer clients, soil samples were sent in but as he said in his paper, *Soil Analysis* before the AAAS a great number of them related, not to the general nature of the soil on the farm but to such things as a sour or scoured patch or some area which yielded unsatisfactory results. There was often no indication of what kind of land they were taken from; land already cleared and cultivated or virgin land. He made no claims for its value, *The examination of soils by this branch of the Department is more in the nature of an investigation than an anlysis, the object being to reveal to the farmer the defects of his soil, such as sourness, stuffness, presence of plant poisons, lack of humus(organic matter), or of lime, etc and to advise the proper treatment, in order that benefit may be derived from the subsequent manuring.*

In 1907 Guthrie gave a paper, *Soil Surveys* before a meeting of AAAS in South Australia, making an argument for a nation-wide soil survey. He said that from the time he had given his paper on soil analysis in 1893 more work had been done on soil fertility and those factors which had brought it about. Disciplines such as soil physics had put to rest the idea that the fertility of the soil could be gauged by the amount of chemical plant food which was present in the soil. He dismissed as misleading the results of chemical analysis based solely on the chemical constituents. The discipline of soil bacteriology had presented a new 'rational view' of soils fertility and his Department had employed a bacteriologist for a few years to enquire into the matter. He then emphasised that it was more important to know the condition of the plant food that was present in the soil rather than know what is immediately available. His introduction was no doubt preaching to the converted but it was a preamble to his appeal to their competency to promote the idea.

What he put before the meeting was not the agenda for the discussion of the advances but the considerable advantage of having a soil survey conducted over the whole of Australia just as other countries were doing. *I think the time has come to bring under the notice of our authorities the importance of carrying out such an investigation locally. Before doing so, it is desirable to have some fairly cut-and-dried scheme to lay before them, and it is the hope of promoting discussion and eliciting an expression of opinion the matter form those competent to deal with it that I venture to lay my views before this meeting.*

Set out in the report of the meeting was the answer to what may have been unreported questions and which would most certainly be the question asked by the bureaucrats: what will it cost and who benefits? His answer was that it was not only for a better knowledge of soils in a given area, their possibilities and needs, but, appealing to their scientific curiosity, it would be a chance to investigate systematically, the numerous unsolved questions associated with the treatment of soils.

As far as the matter of soil analysis for farmers was concerned, the proposed survey would not replace that need. The survey would not replace the individual analysis but it would lessen the work involved if the detailed maps and relevant information were available as the result. The purpose of the survey would be to provide the general characteristics of any given area. The preliminary work would be in the field to classify the types of soil, the dpeth of the subsoil with any variations in their depth to be marked on the map. Samples of the soil would be examined in the laboratory. The surface features, water, vegetation, ridges, and gullies, rocks and heavy timber. It was to be an exercise in the manner of Hilgard.

It was necessary that all States should co-operte to ensure uniformity and continuity in th survey. The work could be carried out by the Commonwealth, with branch bureaux set up in each State. He pointed out that the greatest value of such an organisation would be the survey of land opened up to settlement. In New South Wales, Guthrie, in the course the government's Closer Settlement schemes for opening up new farming subdivisions, had taken part in soil surveys at Myall Creek and Dorrigo, in addition to the 'Barren jack' Irrigation land at Yanco, not then called Murrumbidgee Irrigation Area.

His enthusiasm for the project is evident: *Consider what a mass of information of the first value would be available to the farmer if the whole surface area of one of the States had been mapped out and described in the manner suggested. The nature of the soils and their distribution, their peculiarities, good and bad points, suitability for different crops, the best and the most economic treatment for different purposes would all have been more or less accurately determined and the landowner could at once put the knowledge so gained to practical use without unnecessary waste of time or labour in experimenting.*

Guthrie's remarks echo those of Farrer in the saving of time, money and energy by the farmer if such work were carried out scientifically and systematically. Right from his earliest days in the Department, Guthrie had pressed for a survey of the State; his own first soil survey had been close to home, the County of Cumberland. It had been published in the *Agricultural Gazette* in 1898. There were further reports of other parts but it was not until 1909 that the Department appointed Dr. H. I. Jensen to continue the survey. The report on the South Coast of New South Wales was published in 1910. One other survey was published, *Soils in relation to Geology and Climate*, dealing with the Wagga Experiment Farm, which is an example of the kind of information available in the surveys. It is still a useful document, detailing the tree cover on the Farm at the time, as well as the geology of the site, the soils in different parts of the 3000 acres(1500 hectares). Jensen resigned in 1911 to become the Northern Territory geologist. With Jensen's departure the New South Wales survey came to a halt but in the 1917 annual report of the Department, the need for a soil survey was still being stressed. Shortage of funds from the Appropriation Bills was always put forward for the suspension of the necessary work. The Commonwealth Advisory Council of which Guthrie was member was pressing the Commonwealth Government to carry out the wide survey. Other voices were being heard. A resolution urging that the survey should be started was moved at the Annual Conference of the Farmers and Settlers Association that year. Guthrie was not to see it happen in his lifetime.

DEATH OF FARRER

WILLIAM FARRER had mentioned in his letters that he was feeling tired and lacking energy, but was pushing on with his work. At the beginning of the autumn of 1906 he had been busy working and planning at Lambrigg ahead of his regular visits to the Experiment Farms which he normally took about the end of April. However, about halfway through the month just before Easter, he must have been having chest pains because he wrote a letter to Dr. Richardson his friend and doctor in Queanbeyan, describing his symptoms, angina pectoris.

The day he died was Easter Monday. A number of young relatives were staying with the Farrers at Lambrigg for the Easter holiday. That morning the nieces and nephews all set off for the Annual Picnic and Sports at Tuggeranong, about five miles away, taking with them the picnic baskets Nina Farrer had packed with the results of her baking. They set off on that early autumn day, some on horseback and some in the buckboard, over the Point Hut Crossing to the other side of the Murrumbidgee River. Nina Farrer and her namesake niece, Nina de Salis stayed behind; Farrer was in bed, writing letters and reading his notebooks, planning his next experiments. Mrs Farrer sat quietly by him, reading and keeping him company. At lunch time they had their meal on a tray. Time passed quietly until about the middle of the afternoon when suddenly Farrer was struggling for breath. Both women tried unsuccessfully to help him, but without success and he died in the space of a few minutes. It was about half past three.

Born in the English Spring he died in an Australian autumn.

Hastily, Nina de Salis, Farrer's niece, changed into riding clothes and catching a horse set off for Tuggeranong Sports Ground to seek help, along the same road the others had taken so cheerfully in the morning. Her sisters, immediately she arrived, made ready to go with her and Tom Curley who in all probability would have driven them back in the buckboard. One of Farrer's assistants, perhaps George Norris, rode off to Queanbeyan to fetch the doctor but Dr Richardson was unavailable, so Dr Blackall came instead. Before the girls left they wrote a note to their parents at Soglio at Michelago and sent it off with their young brother Eric. It would have been after four o'clock when he left the sports ground and it was after dark when he reached home. Charlotte de Salis said later that at Michelago they all knew something was wrong when they heard him in the dark passage. Eric was sent to bed because he had to be up early in the morning to catch the Cooma Mail train to take messages to his uncles at Umeralla and then back the other way to Parramatta.

It was eleven o'clock when George de Salis, Nina Farrer's brother set off from Soglio to ride his horse Sergeant to Lambrigg. When he left the moon had not risen and a cold wind was blowing. Riding through the Wright Ranges, it was about two o'clock when he rode into the home paddock at Lambrigg by the light of the newly risen moon. Everybody was still up although Nina Farrer was resting in her room.

The next morning, the 17th of April, George de Salis took matters in hand. Dalhunty was sent off to Queanbeyan to order the coffin and take Dr Blackall back; a letter was sent to St John's Church Rectory in Canberra to the Reverend Gaillard Smith, George de Salis' father -in-law, to come and read the funeral service. By 1906 Gaillard Smith was an old man, having come as rector of St John's in 1853 and was soon to finish his ministry. He sent a message to say that he would if he could. In the event he did not get to Lambrigg.

Nina Farrer knew where Farrer had decided to be buried. It was on the top of the hill that rose behind the house to the west where he could look out across his experimental plots to the hills on the other side of the valley, which had reminded him so much of his native countryside in Westmoreland. George and Nina climbed the hill to decide on the exact location. Having decided the place, George rode over to a neighbour, George White, to ask him to come and dig the grave. White came and with him three others, a man called 'Dan, Old O'Donel and Albert Hannaford', as George lists them in his diary. They began the task after dinner, but about two feet down struck rock. The work had to stop. Dalhunty had to ride to town for explosives. When Mr. Circuitt, the owner of a neighbouring run, called in after tea; he offered the services of one of his employees, an old miner, to come the next morning and blast the rock when the explosives came back. The next day the 18th April, Dan Harris came to begin the blasting.

They had intended having the funeral at midday but by the time the grave was four feet deep and six feet long it was after dark. George de Salis recorded in his diary the succession of events. About dusk, Mr Smith having failed to arrive, everyone gathered on the verandah and George read the first part of the funeral service from the Book of Common Prayer, *I am the resurrection and the life, saith the Lord.* Then the coffin was loaded onto the farm cart and taken along the track up the hill. Nina Farrer walked with it, accompanied by three of her nieces, May, Nina and Emily; Farrer's brothers-in-law, Henry and Willie were there. Some of the long time neighbours and Farrer's employees were also following. By the time the procession had reached the top of the hill it was dark. The wind was up and blowing hard; a light rain was falling while George read, by the light of a lantern, the last part of the funeral service as the coffin was lowered into the grave. The girls then put their flowers on the grave before they all walked back down the hill to the house.

The next day Tom Curley, Farrer's long-time farm workman, put up a fence around the grave and netted it against the rabbits and dogs.

George De Salis was to write of his brother-in-law, *he was a kind-hearted man, with a big mind, and worked very hard for the good of his fellow man, he will long be remembered for his work in improving wheats.*

Complementary to his desire to produce a wheat to be nutritious and palatable in Australia, he took a deep interest in the details of the Federation process. In 1897 he had written a letter to be sent to the Constitution Committee meeting in Adelaide in which he outlined his ideas for the construction of Parliament, one of which was to restrict the number of lawyers allowed to be elected to be 10% of the total members and the other was, that where there were controversial measures passed, then there should be a committee of eminent men to discuss it. During the quest to find a place for the Federal Capital, he was among the men who supported the choice of Queanbeyan and its district. At the time he wrote an article for the Queanbeyan paper which John Gale the editor said was to be unattributed, on the various reasons to support the claims of the Wanniassa Plain.

The grave site was to wait until 1929 when Nina Farrer died, for her to be buried beside her husband.

In 1939, Dr Cora Hind an agricultural journalist with *Winnepeg Free Press* described her visit to Lambrigg:

On June 9, 1936 a perfect Australian winter day of fleecy cloud and golden sunshine, I stood beside the grave of William James Farrer as on a similar day in March I stood beside the tomb of Cecil Rhodes on the Matapo Hills in Rhodesia. Once again Kipling's phrase rose to my lips: 'Great spaces washed with sun'. A glorious prospect of hills, mountains and soft valleys with many homes nestling among the trees; broad fields from which the crops were gathered, cattle grazing by the river; the little village in the distance. A wide difference between the two 'sun-washed' scenes. As wide as the difference between the lives of the two men. Both Englishmen, both graduates from her great universities; both devoted to public service. Rhodes benefited a continent, Farrer made better the world's bread. Much as I have always admired Rhodes the palm goes to Farrer.

MEMORIAL TRUST

I N 1906 GUTHRIE HAD written an biographical article on Farrer for *The Lone Hand,* Norman Lindsay's publication, but the idea of a memorial trust was put forward in the eulogies appearing after Farrer's death in April 1896. In 1906, *The Daily Telegraph* had established the Farrer Memorial Scholarship Fund to promote agricultural education it was nevertheless slow in accumulating financial support from the community, especially the wheat-farming community.

The purpose of the trust money being used for scholarships is evident in Maurice McKeown's letters of the time which reported on the examination for the Scholarship and its results. Although it was announced that all first year students from the Wagga Farm School could sit for the exam in 1908 and the students from Bathurst to sit the following year, it turned out that both Bathurst and Wagga Experiment Farm students sat for the first time in 1909. Under the conditions of the scholarship, all students were encouraged to sit for the Farrer Scholarship, but only four students presented themselves that year: H. C. McLachlan, J. E. Barlow and N. Edwards. To fully prepare them for the exam, they were given two extra lectures by Mr Manager McKeown and Mr Experimentalist McDiarmid, over and above their ordinary lectures. The scholarship winner that year was from Bathurst Experiment Farm but the Wagga students came second, third and fourth. The awards had been determined by a competitive examination and were made on the presentation of a certificate of good conduct and aptitude for agricultural work from the Manager of the Farm. There was also the understanding that the money would be used for second year study.

The candidates were required to sit for a written paper and answer a vive voce exam on the topic, *The Cultivation of Wheat in N. S. W..* In addition they were required to submit a paper which had been prepared at the Farm on Plant breeding and the results of observations amongst the stud and crossbred wheats which grew at the Farm during their residence. The Government wheat experimentalist (at that time George Sutton) would be the sole examiner and the scholarship would be announced by the Minister for Agriculture.

At that time the scholarship was only given to students at the Farm Schools, but in 1911 when the School of Agriculture was initiated at the University of Sydney there was a movement to widen the scope of the scholarships. A five-member Trust Committee had been set up and there was some discussion about the dispersion of the funds. It was decide in that year to endow a scholarship for a person who would be most likely to contribute to the improvement of wheat. This could be *either by original work in the field or in the laboratory, or in the mill or bakehouse.* The Government contribution was to be £50.

In 1911 Guthrie wrote a memorial for Farrer to serve both as a publication by the Department and a promotion for the Trust not only being a biography of Farrer himself but a 'biography' of his work and his wheats, underlining their value.

Both Guthrie and Sutton served as trustees, both promoting the educational and research importance of the grants. They did not always agree. Guthrie moved at the May 1911 meeting that the fund should provide an annual research tenable at the University or one of the Farms. Sutton, at that time about to go to Western Australia, and perhaps speaking for the Minister for agriculture who was he Chairmen of the fund, that he would like to see the money for training at the University some young man who could demonstrate the *powers of observation and originality, either as he was about to enter the university or was already there*. At the next meeting Guthrie's motion was defeated and Sutton's plan was adopted. The first recipient was W. L. Waterhouse who was awarded a part-time scholarship at the University of Sydney to study the effect of superphosphate on the wheat yield under the supervision of Professor Watt. In 1913 the award was given to a student who before he could take up his scholarship had enlisted. The wartime situation forced the postponing of awards until 1920.

In the 1920s there was a marketing thrust to raise £7000 so that the selected candidate could study overseas. However, it was not until 1927 that the trust could award an overseas scholarship. That year it was awarded to E.W.Treloar to study the chemistry of wheat and flour for three years at the University of Minnesota with Hyde Bailey and at Cambridge University with Rowland Biffen.

The decision of the Royal Agricultural Society in 1927 to lend wholehearted support to the Trust's appeal for funds would have heartened Guthrie in the year he died. Sir Samuel Hordern as President of the Council said, *I, personally, have given the sum of £100 to the fund, and trust that the appeal of the trustees will merit the support of the big business interests of our society.*

The funds of the trust were increased very slowly until, after Nina Farrer's death in 1929, it was left £1000 in her will. From 1931 the trust was made an annual grant of £500.

The New South Government would have liked to acquire Lambrigg but problems existed as it was now within the Federal Capital Territory. The Commonwealth Government vetoed the idea that Farrer's wheat plots should still be available for research.

In the 1930's the Trust had support from two men: George Sutton, now Director of Agriculture in Western Australia and Frank Gallagher, the headmaster of Queanbeyan Intermediate High School. The trust was further boosted in 1939 when William Farrer, the wheat-breeders nephew, gave £2000 on condition that the government gave £2000. Failing this the money would be put in trust until it had grown to £4000.

Gallagher felt that his students should have some knowledge of the man, who while living in their own district, contributed so much to the wealth of the country. Gallagher as a dedicated headmaster, also held the view that the students should be involved with a view of inspiring them with the worth of education effort and their right to feel pride in a local man who was a national hero. As a person of some influence in the town he set out to form a committee for the purpose of commemorating within the school the memory of Farrer. The first step was install a plaque in the school but then the project expanded to erect a public memorial. It was a lead up to the Bicentenary year of 1938. It is from the copies of the letters he wrote that the enthusiasm for the project is evident. At the same time a committee in Minyip in the Wimmera were making plans for a memorial of their own to the man who gave them Federation wheat, which saved the Wimmera.

Gallagher set out to organise the construction of the memorial, beginning with a letter to Charlotte de Salis, Nina Farrer's niece, regarding the matter of having an 'heroic head' sculpted for the entry into the town. He had already enquired of Rayner Hoff, one of the State's leading sculptors, as to the cost of such a bust and had been quoted £187, a clerk's annual salary. Rayner Hoff had just completed the sculptures for the Hyde Park War Memorial in Sydney. The total cost which included the architectural part of the memorial was £600 which included a special grant from the Federal Government of £135 and a donation from Farrer's nephew. The people of Minyip were anxious. They had hoped to have their memorial dedicated first but Queanbeyan forestalled them by a few days.

The occasion was a vice-regal occasion. The organisers had hoped for a Royal personage but this did not eventuate. The New South Wales Governor, Sir Alexander Hore-Ruthven came instead.

On this occasion the Prime Minister, Joseph Lyons gave the first Farrer Oration. After a period of slipping obscurity except for those who knew him, this memorial resurrected William Farrer in the public eye. Gallagher wrote to G.W. Walker the Chairman of the Trust in 1937, and said concerning the memorial; *…my objective of making the nation realise the personality and character of Farrer. The monument at Queanbeyan symbolises this, whereas the monument at Minyip in Victoria is a symbol of his work.*

In 1938 Gallagher wrote to the Prime Minister's Department setting out his ideas for the perpetuating of the Farrer Oration: that the orator be decided in December for the oration the following April; arrangements be made for broadcast and that a medal should be presented. In addition he called for a complete biography to be written *as a field of inspiration to Australians, and as a proof that the individual and free activity of one individual in a democratic setting can achieve more than all the vaunted and blatant results of any system of totalitarianism.* Archer Russell produced a biography in 1949. Gallagher left it to Major Casey whom he knew to promote his cause. In the event the Commonwealth thought they had contributed enough, having financed a headstone over Farrer's grave. The suggestion concerning the Oration was put back into the hands of the Farrer Trust, eventuating in the annual Farrer Oration, although not the ABC broadcast as Gallagher had envisaged.

Sutton in meanwhile had been conducting his own campaign of memorialising Farrer. During the 1920s he had corresponded with Nina Farrer adding to his personal knowledge of Farrer. In his position of senior standing in the West Australian Department of Agriculture he had access to newspapers and journals who would publish what he wrote as well as being broadcast. When he had come to Western Australia taken Farrer's philosophy wheat breeding and the value of experiment farms. He had encountered the same obstructions that had bedevilled the New South Wales workers

In 1914, William deSalis, Nina's brother, while living in England wrote to the Farrer Trust and thanked them for continuing for her, William Farrer's annuity from the Trust.

Nina Farrer, after arguments with the Commonwealth Government about the value of the compensation for Lambrigg when the compulsory acquisition of land in the new Australian Capital Territory took place. The matter was finally settled and she was allowed 40acres on a rental of £100 a year for life. When she died in 1929 she left £1000 to the Farrer Memorial Trust.

EPILOGUE

$\mathbf{F}$ARRER'S INITIAL CHALLENGE was to find a solution to the rust problem by finding a rust resistant wheat. However as the 1890s moved along he went searching for a better combination of qualities. He was ultimately, or perhaps had always been, searching for the perfect product but he himself knew that new varieties would keep emerging and that was the basis of his opposition to making a definitive nomenclature of wheat and over which he entered into a controversy with Cobb. Cobb however pushed on published between 1901 and 1905, although it was not revisited until the 1960s.

There were forces which motivated Farrer in his application theory to practice. The 1880s were years of rising prosperity and the accelerated growth of cities. The Italianate and grandiose were on display in townhouse and country homesteads. The idea of Governor Gipps in the 1830s of a pastoralists empire had been replaced by a rising wealthy bourgeois class of business and professional men pushing the growth of industry and technology. By 1889–90, the deSalis family like many others could no longer be counted among the leading land-owners. They had in fact by 1893 finally lost Cuppacombalong and three households had been so contracted in space that they were all living in Lambrigg homestead in what might be assumed as considerably reduced circumstances. Nevertheless, the Farrer hospitality was always evident.

William Farrer without a doubt enjoyed exploring the possibilities which lay in his work as he mulled over his plans for the next round of sowing, growing and harvesting, and finally testing those results which looked the most promising of his possibilities.

His work was not carried on in isolation at Tharwa; the voluminous letters which he sent out and received kept him touch with a wide spectrum of scientists and botanists, from whom he sought advice about projected processes but also about contentious points with which he was wrestling. By 1900 he had become a world figure, respected for his work and ideas.

Except for the flats by the river, the 240 acres of Lambrigg made up of six conditional purchase blocks, were rocky and infested with rabbits. The length of netting and fencing to protect his plots and orchard were recorded by estate agent's valuation schedules

In 1900, Carl Correns, Eric Tschermak and Hugo de Vries brought to light the work of the Austrian monk Mendel, a mathematician, on the variability of inherited characters. In scientific circles of plant and animal breeding it created a sensation.

The next year, in 1901 Farrer was accorded acknowledgment for his wheat breeding work by W.J.Spillman in a paper, *Quantitative studies on transmission of parental characters by hybrid offspring*, which was delivered at the 15th Convention of the Association of American Agricultural Colleges and Experiment Stations.

Spillman said, *Until recently practically all the work was done with ornamental plants only or with those of little economic importance. More recently much has been done with field crops.* Linking Farrer with European and American names such as Rimpau of Germany and Carleton of the US, he continued, *he has attracted general attention to the possibilities of improvement by hybridisation and selection.*

It is well known that types appearing in second and later generations can be fixed by selection to type for several generations, usually five or eight(Federation lasted many more)*It is also important to know just what types to expect.*

Spillman at that time had not been aware of Mendel's Law, which was introduced to American scientists at a conference organised by Liberty Hyde Bailey, a well-respected professor of horticulture.

Farrer corresponded with Spillman who was Professor of Agriculture at the University of Washington in the north-west of the United States. A letter received from Spillman at the beginning of October,1905 is full of enthusiasm for Mendel's law: *I have always contended that this new knowledge was of the highest practical importance.* He went on to outline work he had done by following Mendel, reiterating what he had already written, *you will see I regard Mendel's Law, even with our present limited knowledge, of considerable practical importance, but I do not go as far as Prof. Bateson in stating we may hope in the near future a desired hybrid can be produced with the same degree of certainty that the chemist produces a desired compound in the laboratory, but it does begin to look as if such a thing were possible.*

The American connection which Farrer had established with Blount in the 1880s found a continuation and now, an affirmation of Farrer's reputation in A.H. Danielson's communication. In late June 1905 a letter was sent from the State Agricultural College of Colorado indicating that in the years spanning the time when he had first sought advice from Professor Blount he had continued to exchange samples with the College. The letter had come from the Assistant Agronomy lecturer, A. H. Danielson who was evidently in charge of wheat- breeding. The covering letter was in a sense a pro forma asking if Farrer had received samples which had been sent to him. Added material accompanied the letter outlining the investigations Danielson had been making using Farrer's samples. One had been a variety which he had labelled Nonpareil Elephant which had, in a disastrously rust-infected year before, produced a perfect quality crop of wheat.

Farrer had sent another variety in a packet which he had labelled, *An attempt to make a bald wheat into the qualities of the macaroni.* Danielson found it a great source of interest as he said he was working along the same lines. An inked annotation on the typed letter reads, *We call it 'Farrer's Beardless Durum'.* He assured Farrer that he was he was busy increasing it so that he could make comparisons with the best results of his own developing. Durum, a wheat which had engaged Farrer's attention in spite its dismissal by others from the beginning of his work at Lambrigg, was affirmed as worthwhile in Danielson's letter.

The American had come across a conundrum. For a couple of years he had been growing one of Farrer's crossbreds which had been sent in 1900. On one of them had appeared a *sport* which was *peculiar in many respects and entirely unlike any other.* He went on: *It has a very peculiar clubhead with very strong coarse chaff, with a rather coarse berry. But its most distinguishing feature is its very short, thick straw. It is also fairly early, but I thought at once it might be a very good variety to grow on our very strong soils where grain of any kind falls down.* He was impressed enough with its potential to give 60 pounds of it the year before to a farmer to grown on his very rich soil. It remained upright. The farmer was pleased enough with the result to plant more in the current year. Danielson wanted Farrer to identify it. The laboratory assistants had called it 'Toad' because it looked like a toad on a stick. Officially it was called Colo. No.76

Because in Colorado some farmers were practising farming without irrigation there was a need to find the best crops for it. Durums were in great demand and it was estimated that 10000acres would be planted in 1905. Danielson saw possibilities in Farrer's varieties as they were being bred for that kind of cultivation, running as they were a close second to the durums. He wanted Farrer, if he would, to send him some samples of his latest and best varieties.

Danielson wrote that they were having the same difficulties with their millers as the Australians had had with theirs when it was a matter of milling these extremely hard wheats. One of the local millers had installed special machinery and was producing durum far more successfully then other millers and Danielson and his family were eating bread made with durum flour.

With the ongoing use of durum, Danielson was anxious to develop a beardless durum because it could be used for stock feed in areas of the state where durum will mature where others fail. On farmer had assured Danielson he would pay $1000 for *a start of beardless durum.* The plant breeder considered he had succeeded in his aim as his varieties were now breeding 75% pure.

Finally he asked for Farrer to provide a biography and a photograph. *as your varieties bid to become very important with us.*

In a paper *Improvement of wheat,* published by the United States Department of Agriculture in the Yearbook of 1937 J.Allen Clark, writing about developments in wheat breeding world-wide, included a section he called *Oceania.* Dr Wenholz of the New South wales Department of Agriculture contributed the information on Farrer's work...*One of the world's pioneers in plant-breeding, many thousands, including Federation and Bunyip are now important in U.S.*

In California in1937 the recommended varieties were White Federation, bred by Pridham at Cowra, introduced into the US in 1916 but by 1920 Federation and Bunyip found their way into California.

APPENDIX

T{HE FOLLOWING LIST} is a selection from the collection now housed at Farrer Agricultural High School. They indicate the range of books to which he had access.

An Introduction to Practical Astronomy. Loomis 1873
Animal Products. Simmonds 1889
Annual Report of the Department of Agriculture 1892–93. Brisbane
Conducting a Trigonometrical Survey. Frome 1873
Courses of treatment of Imperfect Digestive. Leared 1882
Digest of Metabolism Experiments. Atwater and Langworthy 1898
Electricity in Agriculture and Horticulture. Lemstrom 1904
Electricity. Ferguson 1867
Fallow and Fodder crops. Wrightson 1889
Farm Drainage. French 1884
Feeding Animals. Stewart 1882
Fermentation. Schutzenberger
Food and Feeding. Thompson 1894
Handbook of Natural philosophy: Electricity, Magnetism, Acoustics. Carey Foster 1874
Herbert's ABC of sense Perception. Eckoff 1896
Heroes of Science: Chemists. Pattison Muir.1873
How to Farm Profitably. Mechi
How to raise fruit, Gregg. 1880
Irrigation for Farm, Garden and Orchard. Stewart 1877
Labour, Leisure and Luxury. Alex Wylie. 1887
Les Engrais Chimiques. Ville. 1867
Lessons in Elementary Physiology. Thomas Huxley. 1892
Natural and Artificial Methods of Ventilation 1899
Peat and its Uses. Johnson. 1866
Practical Poultry -keeping. Wright 1889
Principles of Political Economy. John Stuart Mill
Quain's Elements of Anatomy. Sharpey 1867
Small Fruit Culturist. Fuller,1885
The Agriculture of the United Provinces. Moreland 1904
The Country Industries of Victoria: Viticulture 1894
The Rose Amateurs Guide. Rivers 1877
The School of Chemical Manures. Ville, translated Fesquel 1872
The Science of Politics. Sheldon Amos 1883

The Shepherd's manual. Stewart 1879
Transverse Tables. Boileau 1872
University of Nebraska. 7th Report 1894
USDA Handbook of Experiment Station Work 1893

BIBLIOGRAPHY

Books

Agricultural Gazette, 1890–1913

Amundson, Ronald and Yaalon, Dan. H, Hilgard.E.W. and Powell. J.W.: Efforts for a joint Agricultural and Geological Survey. reprint for the *Soil Science Society of America Journal*

Moyal, Ann. Nineteenth century Australian Scientists.

Atkinson. James, An Account of Agriculture and grazing in New South Wales

Back o'Bourke

Blanchard Frieda Cobb: Nathan A. Cobb, Botaniser and Zoologist. Pioneer Scientist in Australia.The Asa Gray Bulletin 1957

Brent of Bin Bin, Up the Country

Burkes Landed Gentry

Bust, Boom and Bust: some Reminiscences of wheat and wheat breeding in Australia/compiled by F.E.Stanton 1984

Buxton, G.L.. The Riverina, 1851–1891

Callaghan and Millington The Wheat Industry in Australia

Campbell, W. S. An historical sketch of William Farrer's work in connection with his improvements of wheats for Australian conditions. 1912

Cantlon Maurice. Homesteads of New South Wales.

Clark J. Allen, Senior Agronomist USDA. Improvement of wheat; United States Department of Agriculture, Yearbook Separate No.1570, 1937

Crookes, William. The Wheat Problem 1917

Farrer,Keith. A Settlement Amply Supplied

Farrer,Keith. A Settlement amply supplied

Flannery,T. 1788. Watkins Tench 1996.

Franklin Miles, All that Swagger

Guthrie.F.B. Wheat and Flour Investigations. Department of Agriculture N.S.W. Science Bulletin 7.

H.F.Roberts. Plant Hybridization before Mendel. Princeton University press 1929

Jenkins The Green Revolution: Genes and the Cold War

Jensen H.I., Soils in relation to geology and climate. 1912

Journal of proceedings of the Royal Society of New South Wales. 1903,1927

Large E.C. Advance the Fungi

Macindoe and Walkden Brown: Wheat Breeding and Varieties in Australia 1958

Mary-Ann Tylered Kate Gibbs. The Adventurous memories of a Gold Diggeress, 1841–1909

Musgrave, Sarah. The Wayback
Mylea P.J. In the Service of Agriculture. A centennial history of the New South Wales Department of Agriculture, 1890–1990
Proceedings of the Congress of Agriculture, Paris 1900
Sayre Richard, The Art in Phytopathology: Portfolio of Nathan A. Cobb nemotologist
Spillman W.J., Quantitative Analysis of wheat, 1901
Sutton. George, Come the Harvest:half a Century of Agriculture in Western Australia
The Commonwealth of Science: ANZAAS and the Scientific enterprise in Australasia, 1888–1988.ed.Macleod. Roy

Periodicals
The Australasian Sketcher 1882,1890.
The Daily Telegraph, 1905–1906; 1927
The Garden and Field, 1893
The Queenslander 1882
The Sydney Mail 1893–1905;1934
The Sydney Morning Herald 1913
The Victorian Naturalist
The Victorian Naturalist
The Wagga Wagga Express 1892, 1897
Wagga Wagga Advertiser 1890–1905

Collected Papers
de Salis papers, National Archives of Australia.
deSalis Collection, National Library of Australia
Farrer papers, Mitchell Library
Farrer papers,Kendal Public Record Office
Farrer papers,National Archives of Australia
George Sutton Collection, Battye Library.
Guthrie material, ex Colin Wrigley
Nathan Cobb material. Dorothy Blanchard.
Shelton Collection letters, William Farrer to Professor Galloway and Mark Carleton, Department of Agriculture
Sherry Morris Collection
W.S. Clarke letters, Mitchell Library
Wagga Experiment Farm letterbooks, 1894–1906

ENDNOTES

1. Rose Hill as a name still exists but is now included in the City of Parramatta
2. Canterbury Vale now the inner suburb of Sydney, Canterbury
3. An acre is the equivalent of 0.41 hectares.
4. Creeping wheat is a wheat with a prostrate growth until the stalks begin to grow vertically holding the seed head.
5. An average days wages for a competent workman at the time was 5 shillings.
6. This extract was quoted by Archer Russell, but I have been unable to find the original book.
7. `Mr Downes MLA wrote a eulogy for Farrer in the *Sydney Morning Herald*
8. 'Democracy's College in the Centennial State. A History of Colorado State University'. By James Hansen II
9. In July 1896 Farrer wrote to Lowrie at Roseworthy Agricultural College, *I am distributing as widely as I can for trial in different parts of Australia and the reports I have received I record carefully. I have already found out that the sorts that thrive best in Queensland are not the kind that do well here, also that the same wheats appear to suit widely differing climates of the coast and the interior of Queensland*
10. Professor Shelton had recently arrived from the United States to take up a position with the Queensland Department of Agriculture.
11. A loaf of bread cost 3pence at the time
12. See Chapter: Visitors at Lambrigg
13. Monterey pines, natives of California
14. Aleurone is the albuminous or proteinous substance found in granules in seeds
15. Macindoe and Walkden Brown, *Wheat breeding and varieties in Australia.* 1968
16. Aimé Girard in 1888 and 1889 investigated the prevention of potato blight.
17. Professor Sitensky was very famous in Bohemia as the publisher of 4 volume Encyclopaedia of Agriculture. He established the Research Station for Agriculture and Botany. He was an agronomist. His research station was famous for the breeding of forage crops, especially clover. The breeding of cereals was also carried out
18. Macindoe and Walkden Brown, *Wheat breeding and varieties in Australia.* 1968
19. Valder was the Farm Manager from 1894–1897
20. Only departmental officers were permanently employed
21. there is some criticism about the validity of this cross